T0210690

The Stumbling Progress of 20th Century Science

Lars Jaeger

The Stumbling Progress of 20th Century Science

How Crises and Great Minds Have Shaped Our Modern World

 Springer

Lars Jaeger
Baar, Zug, Switzerland

ISBN 978-3-031-09617-4   ISBN 978-3-031-09618-1   (eBook)
https://doi.org/10.1007/978-3-031-09618-1

This Springer imprint is published by the registered company Springer Nature Switzerland AG
The registered company address is: Gewerbestrasse 11, 6330 Cham, Switzerland

Dedicated to my daughter Anika Mai Jaeger

Prologue

"Science doesn't always go forwards. It's a bit like doing a Rubik's cube. You sometimes have to make more of a mess with a Rubik's cube before you can get it to go right."
Jocelyn Bell Burnell, *Radio Astronomer*[1]

By the end of the nineteenth century, almost all scientists were convinced that they had thoroughly understood the laws of nature and thus everything about the very essence and depth of the world. Newton's laws were regarded as an eternally valid world formula, and the recent findings in the fields of magnetism and electrodynamics seemed to round the picture off beautifully. With this attitude, it happened that, when the young Max Planck asked one of his teachers in the 1870s whether he should study physics, he was given the answer that there was not much more to discover in the field. Fortunately, Planck did not listen to this advice.

Werner Heisenberg, who revolutionised physics two generations later in the 1920s, was already one step ahead:

"Few know how much you need to know to know how little you know."[2]

---

[1] Jocelyn Bell Burnell discovered the class of stars known as pulsars, but the Nobel Prize for this discovery was awarded to her male colleagues.

[2] Source: *Fliegende Blätter, humorous weekly German paper, 1845–1944. Heisenberg was probably inspired here by Socrates, whose sentence "I know as a non-knowing person" or "I know that I do not know" is often misquoted as "I know that I know nothing".*

Even knowledge that had been so painstakingly gained and was believed to be certain proved in the end to be unexpectedly volatile. What earlier generations of researchers believed to be absolutely true is, in the vast majority of cases, no longer true for us today. Scientists have learned this lesson. Today, they explicitly assume that knowledge can only ever be temporarily correct. What is valid today can turn out to be wrong at any time.

This turn away from eternal claims to truth began at the end of the nineteenth century and triggered a comprehensive crisis in the sciences. In the eighty years from 1870 to 1950, a period that is a blink of an eye in the history of humankind and covers not much more than a human lifetime, there was what is probably the greatest revolution in thinking of all time. It was far more significant than the paradigm shifts of the Renaissance and the Enlightenment (even if one hears little about it in school history books). This crisis in the sciences was accompanied by two world wars, the downfall of traditional social orders, and a reorganisation of the world.

Where suffering is great, salvation is not far away. From the end of the nineteenth century and in the first half of the twentieth century, a number of scientific geniuses of breathtaking creativity were at work, ultimately leading the sciences out of this crisis. This book describes their contributions and follows science on its exciting and bizarre journey into the modern age. Along the way, we shall meet, among others, the mathematical and physical genius of James Clerk Maxwell, the intellectual giants Georg Cantor and Ludwig Boltzmann, engaged in serious psychological struggles, Charles Darwin, who was so moved by questions of faith as well as science, the reluctant revolutionary Max Planck, the Swiss revolutionary Albert Einstein, numerous ingenious and youthful physicists gathered around Niels Bohr, who overturned the world of physics for good at the age of not much more than 20, and last but not least the mathematical geniuses John von Neumann, Kurt Gödel, Alan Turing, and Emmy Noether, whose revolutionary thinking would not even stop at the basic principles of logic.

This book is divided into two parts. Chapters 1–5 describe a first phase that can be roughly associated with the period between 1870 and 1925, in which developments took place almost simultaneously in physics, mathematics, biology, and even psychology that led to the deepest crises in those disciplines. Chapters 6–12 discuss those same sciences in the subsequent years up to 1950, which brought about the transition to modernity. This period also marks the decisive turn from a science oriented towards theory and philosophy to its present more practical and application-oriented approach.

Baar, Switzerland   Lars Jaeger

Contents

Part I

The Great Confusion

With Newton's mechanics, the physicists of the 18th century had at their disposal for the first time a fundamental theory of nature based on natural laws. They no longer had to use their faith or metaphysical speculation to explain the objects of their experience of nature. They only had to solve mathematical equations. Newton's laws provided them with the necessary world formula to do so. The resulting explosion of scientific knowledge had its due effect on the mindset of European societies in the 18th century: the Enlightenment was the intellectual offspring of Newton's physics. And with the technological possibilities arising from the new knowledge of nature, an ever stronger optimism for the future developed in the course of the 19th century.

During this time, natural scientists also set their sights on phenomena that had remained unexplained since antiquity: chemical transformations, heat, electricity, and magnetism. In doing so, they recognised that they had to grasp the behaviour of the smallest particles and made an important assumption: the laws of physics should also apply to the microcosm, i.e., to processes beyond direct observation, just as they apply to the macrocosm. In particular, the same determinism should apply here as in Newton's laws and those of electricity and magnetism. But this assumption contained philosophical explosives: the closer the physicists tried to grasp the properties of the smallest particles, the clearer it became that their behaviour did not follow the known laws of physics. They discovered something here for which there was no place at all in Newton's physics, but which seemed to play a fundamental role in events in the microcosm: chance.

With Charles Darwin's theory, biologists also found themselves exposed to a revolutionary new paradigm: suddenly, man was no longer a creature of God, but a product of nature. And Darwin also had to make a decisive assumption in his theory of evolution: as in physics, tiny particles had to act in biology, behaving according to the rules of chance. Thus, at the end of the 19th century, the ancient enigma of the smallest particles entered the stage of scientific thought with full force. At the end of the 19th century, physicists had only just got used to the idea that their theories would soon enable them to understand the world completely and in all its details, and that they would only have to worry about a few remaining minor problems, when their mental edifice collapsed.

However, it was not only the sciences that were hit by crisis. Mathematics also fell into a deep one at the end of the 19th century due to its now emerging inner contradictions and it had to fight a no less difficult battle for its integrity in the first decades of the 20th century. Revolutionary discoveries were no less prevalent in the scientific investigation of our own minds: psychology now described our mind as an uncontrollable beast. After it had shown its ugliest grimace yet in the First World War, a new age was already looming on the horizon that would make science the plaything of political interests: totalitarianism.

1

Newton's World Formula that Wasn't One—How the Speed of Light Shook Up Classical Physics

Throughout almost all of human history, there was little doubt that God alone—or an ensemble of gods—determined what happened on earth. In everything that humans observed or encountered, his will was revealed. If an arrow hit its target, it was only because God had willed it to do so. The laws of planetary motion discovered by Copernicus and Kepler were also regarded as proof of divine omnipotence. Galileo Galilei had already suspected that the "book of nature was written in the language of mathematics", but until 1687 no one had really tried to open this book and take a look at what would be a completely different way of explaining the world. In that year, Isaac Newton published his work "The Mathematical Principles of Natural Philosophy" (Latin original: Philosophiae Naturalis Principia Mathematica). He was not the first person to link his observations with mathematical calculations, but never before had there been a self-contained scientific theory that provided a rational and causal explanation for almost all of the natural phenomena known at the time. In order to cope with this gargantuan task, Newton had had to develop completely new ways of calculating. It was only with the integral and differential calculus that it became possible to calculate with mathematical precision the motions of the objects around us and the forces to which they are subject. Just as Galileo had predicted, mathematics actually proved to be the decisive tool for describing nature.

Philosophy of nature: The attempt to explain the processes in nature and to form a unified view of the world.

© The Author(s), under exclusive license to Springer Nature Switzerland AG 2022
L. Jaeger, *The Stumbling Progress of 20th Century Science*,
https://doi.org/10.1007/978-3-031-09618-1_1

Newton had achieved this breakthrough not least because he had discovered gravity as a force. All other forces known at the time acted *directly* from body to body. For example, a cart could only be pulled by a horse if both were connected by a drawbar. But gravity also works when the bodies do not touch each other at all. When Newton understood that the force that keeps the planets in their orbits is the same as the force of gravity that makes an apple fall to Earth, he was able to formulate his universal law of gravitation. With this law and the three basic laws of mechanics, almost all phenomena known at the time could be rationally explained and even predicted. The trajectory of a stone thrown through the air, the way a pulley makes work easier, the tides, even planetary orbits and the recurrence of comets had now become comprehensible. Now it was only a small step to an explanation of the world from the ground up that no longer had to resort to the will of a god or other transcendent causes.

The Last Magician

Newton did not adopt this approach throughout his whole life. His mathematical laws were purely rational, but he himself was not a very rational person. Deeply religious and with a strong inclination towards the occult, he spent a lot of time searching the Bible for hidden clues and secrets. When in 1650 the Irish archbishop James Ussher thought he had proved through the most exact and careful study of the Bible that the day of creation was 23 October 4004 BC, Newton checked this date against the astronomical constellations and came to the conclusion that the world had to be 534 years younger than Ussher had calculated.

As an ardent follower of alchemy, Newton believed in a spiritually active substance called aether which permeates every solid substance and also circulates between the centre of the earth and the celestial bodies. In his manuscripts, he repeatedly referred to the aether, which in his view, as the originator of gravity, should push all matter towards the centre of the Earth or Sun. It is paradoxical that Newton's deep belief in alchemy and in the all-encompassing work of divine, alchemical, and astral forces in nature led him to develop the idea of an invisible force of gravity, which pulls things to the Earth's surface and also keeps the other planets in their orbits. The fact that it was Newton, a convinced secret ally and occultist, who helped modern, strictly rational science to make its decisive breakthrough is one of the strangest twists in the history of science.

Occultism: Worldview in which only a few people are given access to hidden, mystical powers. The occult sciences include alchemy and astrology.

On 17 July 1946, a commemorative speech was read out that at Trinity College, Cambridge, Newton's former place of study, on the occasion of the tercentenary of Newton's death. It had been written by the famous British economist John Maynard Keynes, who had died only a few weeks earlier:

In the 18th century and ever since, Newton was thought to be the first and greatest representative of the modern scientific age (...) But Newton was not the first of the Age of Enlightenment. He was the last of the magicians, the last of the Babylonians and Sumerians, the last great mind to look upon the visible and intellectual world with the same eyes as those who began to build our intellectual heritage scarcely less than a thousand years earlier.

Newton's religious faith had been unshakeable, but his laws were now out and about in the world and weakening the power of religions everywhere. With the explanations they gave, the world was no longer a place where the inexplicable will of God made incomprehensible things happen. People began to understand nature—including man himself—as a machine that follows universally valid and comprehensible laws and that can be mathematically calculated and predicted with the help of Newton's equations. Newton became the hero and idol of his age.

The influence of the new thinking on European societies was enormous. For many centuries, countless scholars had struggled to describe and understand what was happening in nature. Especially since the Renaissance, they had increasingly found the courage to question religious dogmas. However, since they had had to fear persecution, they had corresponded only amongst themselves. People outside these scholarly circles—from aristocrats to peasants—had been excluded from such ways of thinking, and their lives and experience had remained the same. But this changed with Isaac Newton. The great philosopher Voltaire had played an important role in this. He had wanted Newton's works to be translated into French (but he could not do it himself, so instead his long term lover, the brilliant female mathematician, Émilie du Châtelet, did it between 1745 and 1749) and spread his enthusiasm for Newton's laws in his native France. From there, the new thinking conquered the whole of Europe. There were heated discussions in the coffee houses and private salons of Paris, Berlin, and London, but also around the table in country villages. Freed from the religious stranglehold on interpretation, people now began to question the nature of political power. If it was not God, who or what would really determine the position and rights of every human being? Who or what would confer political power? Newton and his laws opened the door to the Age of Enlightenment, in which the new rationality also revolutionised social and political thought. The American Declaration of Independence of 1776 and the French Revolution of 1789 would probably not have happened without the Enlightenment, fuelled as it had been by Newton.

> **Enlightenment:** From the end of the seventeenth century, a new way of thinking spread across Europe that relied on human intellect and reason and sought to overcome ideology, superstition, and prejudice.

Newton's edifice of natural laws was not perfect, however. Some phenomena that scholars dealt with in the ensuing period could not be explained by his mathematics:

1. Chemical conversions
2. Heat and cold
3. Electricity and magnetism.

But the optimism for the future was great. The natural scientists were convinced that they only had to supplement Newton's laws with a few more formulae. It seemed to be only a matter of time before humans would be able to understand and calculate everything without exception.

Blank Spots on the Map of Science

The mystical activities of the alchemists had already amassed some knowledge about chemical transformations, but the nature of even the most everyday substances such as water and air was still completely unknown to Newton's contemporaries. For the naturalists, the gateway to understanding chemical reactions was combustion. It was obvious that air had to play an important role in this process, because without a supply of air, combustion never gets under way.

- A significant step was taken in 1781 by the Englishman Henry Cavendish. From sulphuric acid, iron, and zinc, he obtained a gas he called "combustible air", known today as hydrogen. His discovery that this gas produced pure water when burned was a shock, because ancient doctrine had classified water as one of the four basic substances. But if water consists of "combustible air" and oxygen, which was also isolated shortly afterwards, then there had to be more elements than just air, water, fire, and earth. A millennia-old certainty had turned out to be impossible.

- For a long time, natural scientists were convinced that every combustible material contains an invisible substance called phlogiston (from the Greek *phlogistós*: "burnt"). It seemed logical: as soon as the phlogiston escaped from burning wood, for example, ash remained. The chemist Antoine-Laurent Lavoisier made a decisive breakthrough in 1772. In a series of experiments, he burned substances such as phosphorus and lead, even diamonds, and thanks to his elaborate laboratory equipment and the use of precision scales, he was able to measure the initial and final products of these reactions very precisely. He found that the total weight of all products remained constant before and after a combustion. So no material was lost, and nothing was added. But where was the phlogiston? After this discovery, the phlogiston theory would not be able to hold on for long.

- Soon afterwards, Joseph-Louis Proust discovered that the chemical compounds he studied contained integer ratios of the elements involved in the reaction processes, e.g., 3:2 or 2:1. Another Frenchman, Joseph Louis Gay-Lussac, showed in 1808 that this principle also applied to gases. These were the first indications that tiny, indivisible particles were involved in chemical processes.

1. **Chemical conversions**
2. Heat and cold
3. Electricity and magnetism.

Lavoisier did not live to see the successes of his compatriots. He became a victim of the French Revolution. Having previously held important offices in the service of the king, he was arrested, but his fate was sealed by a completely different circumstance. A few years earlier, he had denied the revolutionary leader Marat access to the French scientific elite. Marat now saw a welcome opportunity for revenge. Lavoisier was executed by guillotine in 1794. Even today, the great scientific nation of France still grieves that one of its most brilliant minds was executed in this way.

Steam Flasks and Stills

Like the specific chemical conversion of substances, the mastery of the phenomena of **heat and cold was** also of great practical importance. In particular, the development of steam engines in the late eighteenth century required a deeper understanding of the conversion of heat into mechanical energy that was lacking. Even then, this knowledge could have decisively advanced technical processes and old crafts, such as the distilling of whisky. But Newton's laws did not allow a mathematical approach here either. For a long time, it was believed that, rather as in the phlogiston theory, heat was also a substance contained in all materials. Even Lavoisier, who had brought down the phlogiston theory in chemistry, was an ardent advocate of the view that heat was a substance. He referred to it as "calorique".

- The American Benjamin Thompson studied the relationship between friction and heat. In 1798, he observed how the barrels were drilled into cast bronze cannons in the royal gun foundry in Munich. This was hard work that had to be laboriously carried out using metal drills. He noticed that the bronze heated up again and again with each work step, suggesting that friction was an inexhaustible source of heat. However, a "heat material" would have had to run out at some point after dozens of repetitions. It was thus clear that Lavoisier's caloric theory had to be wrong. Thompson described heat as a form of energy. However, he was too little versed in the sciences to make connections with other areas of physics, especially mechanics.
- Physicists had realised in the eighteenth century that gases could be compressed in a sealed container. This suggested that a gas might consist of particles that whirl around wildly like little balls. External pressure could then reduce the space available to them. In 1811, the Italian Amadeo Avogadro drew the conclusion from his measurements that equal volumes of gas at constant temperature and pressure contain the same number of particles. This resulted in a universal law relating volume, pressure, and temperature, i.e., a law that was the same for all gases. Newton would have been delighted—everything was still in accordance with his laws.
- About 50 years after Thompson, the British physicist James Joule succeeded in developing a mathematical theory for the conversion of mechanical energy into heat energy. In it, he established that heat is related to the motions of the smallest particles.

1. Chemical conversions
2. **Heat and cold**
3. Electricity and magnetism.

In both fields—chemical transformations and the phenomenon of heat and cold—the natural scientists had reached a similar point. There was still nothing to say that Newton's laws were not also valid in these areas. Indeed, they had already found some connections that showed analogies to Newton's laws. But if they wanted to understand chemical transformations and also heat and cold from the bottom up, they realised they would have to deal with the smallest particles.

Ancient thinkers like Democritus and Aristotle had already thought about the building blocks of the world. At the beginning of the nineteenth century, the smallest, indivisible things left the field of philosophy and became the subject of concrete scientific considerations and experiments. However, atoms could not be observed directly; their existence and the laws of their behaviour could only be deduced indirectly with the help of models and hypotheses. For the English naturalist John Dalton, the integer ratios of the initial and final substances of chemical reactions were concrete indications that all matter is actually composed of the smallest, no longer divisible particles. In honour of Democritus, Dalton called them atoms.

To explain the large number of basic substances found so far—among them hydrogen, oxygen, sulphur, and iron—he created the idea of chemical elements, each consisting of a certain kind of atoms. Their distinguishing feature, however, was not to be their external shape or colour, as the Greek thinkers had imagined, but their weight. Dalton gave the atom of hydrogen the weight one, and determined the weight of the atoms of all the other elements as an integer multiple of it. According to this scheme, he classified 21 different elements. Today, chemistry knows far more than a hundred.

Now the natural scientists made an understandable but fatal assumption: they assumed that Newton's laws, which in their experience perfectly described the processes in the visible and tangible macrocosm, should also apply to the microcosm. For them, it was self-evident that two atoms should behave according to Newton's laws just as two marbles do. They were convinced that they only had to find a way to apply Newtonian mechanics to the smallest particles in order to be able to calculate what happens in chemical reactions and thermal phenomena. Then they would also get to the bottom of previously unsolved mysteries, such as which forces hold the atoms together in their compounds.

There was only one catch: with the third physical phenomenon that was not covered by Newton's laws—electricity and magnetism—one quickly reached the limits of what could be explained. There was no evidence of the involvement of the smallest particles and no one had any idea whatsoever of how these phenomena came about.

The Twitching of Dead Frogs

Unlike chemical transformations and thermodynamics, **electricity and magnetism** initially played no role in industrial processes. On the contrary, an air of mystery surrounded these phenomena. At eighteenth century social events, they served to entertain the guests, making each other's hair stand on end.

- The Greeks had already recognised that amber rubbed against a cloth attracts certain objects. This fact also gave the phenomenon of electrical attraction its name: "amber" means *electron* in Greek.
- In the seventeenth century, the German natural scientist Otto von Guericke showed that electricity can be transferred from one body to another. He discovered two forms of electricity, one attractive and one repulsive.
- Like the phlogiston theory, the first theory of electricity postulated a special substance that was responsible for all electrical effects. A convinced supporter of this theory was Benjamin Franklin, who became famous in the struggle for American independence and was also the most important American researcher of the eighteenth century. He believed that too much of the "electric phlogiston" makes the body "positive", too little of it "negative". Because small flashes of lightning could be observed in the laboratory when electricity was discharged, he had the idea that the lightning that occurred during storms could also be related to electricity. He devised a dangerous experiment: according to his description, during a thunderstorm, a kite was allowed to rise into the air on a wet hemp string (which would conduct electricity), with a metallic object attached to the lower end. Franklin recounts that, when lightning flashed across the sky, he put his hand to the metal and received a shock. Presumably, Franklin was only describing a thought experiment; many naturalists who actually imitated his experimental setup lost their lives.
- In 1785, the Frenchman Charles-Augustin de Coulomb measured the force with which two balls repel or attract each other, having first applied a certain amount of electric charge to each. It turned out that this force

depended directly on the amount of charge on the balls. What's more, at three times the distance from each other, the force dropped to one ninth. This relationship, known today as "Coulomb's law", showed the same regularity as Newton's law of gravitation! Electricity was therefore not as mystical as first assumed. It also functioned according to simple, easily understandable laws, just as Newton had formulated them.

1. Chemical conversions
2. Heat and cold
3. **Electricity and magnetism.**

Until 1780, natural scientists knew electricity only in the form of stationary charges with which certain materials, especially metals, can be charged and which discharge abruptly, often sparking, when in contact or close to other bodies. That year, however, Luigi Galvani noticed muscle contractions in the legs of dead frogs when they came in close proximity to electric charges. He knew Benjamin Franklin's theory that thunderbolts were nothing more than electrical discharges. So he hung the dead frog legs in front of a metal net and waited for a thunderstorm. And indeed, when there was lightning, the frog legs twitched. In further experiments, without knowing it, he developed electric circuits from two different metals, using the salt water contained in the frog's leg as an electrolyte.

Galvani interpreted the twitching of the frog's legs as a residue of animal powers. Another Italian scientist and contemporary of Galvani, Alessandro Volta, drew quite different conclusions from the frog leg experiment. He concluded that the effect must be due to the arrangement of the different metals. To investigate the phenomenon further, he took a silver coin and a piece of tin foil, connected them with a copper wire and placed them under or over his tongue. Maybe his tongue would twitch just like the frog's legs, he thought, because after all it was a muscle too. It did not twitch, but Volta felt a tingling sensation, followed by a sour taste. He concluded that his sensations were related to flowing electricity, and that metals not only conduct electric charges but can also produce them. To enhance the effect of flowing electricity, he constructed a stack of alternating layers of metals and salt water solutions around the year 1800. With this, Volta had built the first electric battery! Such an arrangement could produce amazing things, even light, if the two metal end layers were connected with a thin wire of suitable material.

But what exactly produced the electric current in Volta's experiments? The young English scientist Humphry Davy suspected that chemical reactions caused the electric current. Wasn't it then also possible that electric current could cause chemical reactions in reverse? In a sensational experiment, he put the two metallic ends of Volta's arrangement into a potash lye (a water solution with potassium hydroxide). A gas then formed at one end of the cell, which quickly escaped, and at the other end a metal that looked like mercury but burned quickly and explosively on contact with air. The gas was hydrogen and the hitherto unknown, highly reactive metal was potassium. It was now clear that electric current was indeed capable of separating chemical compounds into their components. Davy concluded that it is also electrical forces that hold the components of chemical compounds together. This provided only a first outline of the new force, which was still missing from Newton's explanation of the world; no uniform picture of it was yet available. But the mystery of electricity had become a calculable phenomenon. Efforts were now directed more towards the technical implementation of ideas about what electric current could be good for. In 1802, Davy had experimented with strips of charcoal to generate light by means of an electric circuit. Later, paper, bamboo, and platinum were tried out—with limited success. It would take over a hundred years before Thomas Alva Edison's incandescent lamp conquered the world.

Until well into the nineteenth century, it was possible to assume that Newton's world view only needed to be supplemented in order to solve the last riddles of nature. Regarding all the mysterious forces that the father of classical physics had not been able to explain, scientists were well on the way to integrating them into his explanation of the world. Only one last phenomenon still resisted this integration: magnetism. A student of Davy's, Michael Faraday, succeeded in connecting this force with the new electrical theory. This should have been the keystone that would complete Newton's edifice. But things turned out very differently than expected. Instead of completing and crowning the tireless work of countless natural scientists who had dedicated their lives to science, this very stone brought everything crashing down.

Magnetic Forces and Magical Thinking

Like electricity, magnetism has been known since antiquity. The first statements about its property of attracting certain things came from the pre-Socratic Thales of Miletus, who in the early sixth century BC examined magnetic stones from the area of Magnesia in Thessaly. It was also known that magnetic splinters orientate themselves in a north–south direction. In

China, floating compass needles were in use from the end of the eleventh century and in Europe, magnetic compasses with suspended needles became established from the thirteenth century. Despite these useful applications, the subject of magnetism retained its aura of mystery. The reason for this is probably that, unlike chemical reactions, heat and cold, and of course electricity, a person cannot directly experience magnetism with any of his or her senses, and so it remained shrouded in mystery to a considerable degree.

- Albertus Magnus, the great universal scholar of the thirteenth century, wrote in his book "On Minerals" that the magnet's ability to attract or repel is also useful in marriage: "The husband (may) place a magnet under his wife's head pillow. She will embrace him ardently if she is faithful to him, whereas she will be flung out of bed in the middle of her sleep if she has broken the marriage".
- The English physician and physicist William Gilbert, who described the Earth as a great magnet as early as 1600, imagined magnetism as "the soul of the Earth".
- In early modern medicine, magnets were considered a symbol for the hidden healing powers of nature, and magnetism played a central role as an occult healing instrument.
- In the eighteenth century, Franz Anton Mesmer founded the theory of "animal magnetism" (mesmerism) and described special methods of pain relief using magnets.
- From the middle of the nineteenth century, magnets played an important role in psychological suggestion therapy. "Magnetic field therapy," in which a patient is exposed to an external magnetic field, is still used as a (from a scientific point of view questionable) method of treatment in alternative medicine.

How could it be that magnetism was still seen as a magical force in the times of the Enlightenment and to some extent even up to the present day? In the nineteenth century, the so-called philosophy of life and Romanticism formed as counter-reactions to the strict rationalism that Newton's laws had carried into philosophy and science. The idea of an animal magnetism as well as an animal electricity fitted very well into this world view, which was characterised by belief in ghosts and miracles, and which had also given rise to the idea that polar opposites in constant struggle with each other give rise to a natural whole in the form of a force field.

Philosophy of life: This attempt to explain the world does not rely on reason, but on intuition and feelings. Nature and spirit are understood as a "natural whole" that can by no means be broken down into the smallest particles and understood rationally.

It thus came about that in the first half of the nineteenth century the development of theories about electricity and magnetism was embedded in a larger philosophical debate. Non-physicists such as Goethe and Hegel also participated in the discussion about nature. Schelling's natural philosophy, Schopenhauer's metaphysics, and Fichte's thought were strongly influenced by electrical and magnetic phenomena. One of the few scholars at the time who dealt with magnetism on a rational level was Carl Friedrich Gauss, one of the greatest mathematicians of all time. To carry out his comprehensive and reliable measurements, he invented the magnetometer, which measures the direction and strength of a magnetic field. In the end, the anti-rational thinking of the Romantic school proved to be a temporary phenomenon that did not prevent the scientific study of magnetism.

Faraday: From Magic to Science

Until the nineteenth century, physicists knew even less about magnetism than they did about electricity. The fact that the two are connected was a chance discovery. In 1819, the Danish physicist Hans Christian Ørsted did some experiments with electric current in a lecture at the University of Copenhagen. A magnetic needle was placed near a wire through which an electric current was flowing. When Ørsted switched on the current, the needle stuck out perpendicular to the wire. Puzzled, Ørsted turned the apparatus around. But the needle adapted itself and, as soon as the current began to flow, struck out again at right angles to the wire. Other researchers, such as the Italian Gian Domenico Romagnosi, had already made similar observations. It was Ørsted who first recognised the significance of his observation. As a convinced follower of natural philosophy, he had already suspected years before that electrical and magnetic forces act on each other, because he also had a deep belief that all forces could be transformed into each other. Now he had the proof: magnetic and electric forces could no longer be studied separately; they had to be interpreted together.

Natural philosophers were in high spirits: their view that all the forces of nature are interconnected or even just different expressions of a single fundamental force had gained further support. This view, typical of the

Romantic period, was also held throughout his life by a young man who entered the annals of physics history at precisely this time. His origins by no means predestined Michael Faraday to become the most important experimental physicist of the nineteenth century. At the age of twelve, a craftsman's son with nine siblings, he left school with no prospect of ever receiving a university education. But as an apprentice to a bookbinder, he got to see numerous books, and it was especially those with scientific content that interested him. One day a customer gave him tickets to a series of lectures by the man who had shown that electric current could cause chemical reactions: Humphry Davy. Faraday was thrilled. He sent his detailed notes on this lecture to Davy along with an application for an assistantship. It was one of the fateful moments in the history of science—Davy actually gave the young journeyman bookbinder a job as an assistant in his office.

It had long been known that iron filings scattered around a magnet arrange themselves in a characteristic way around it. Faraday was the first to describe the resulting patterns as lines of force of a magnetic field. He also recognised that a magnetic field forms around a wire in which an electric current flows. Thanks to him, a completely new and soon central concept entered physics: the concept of the field.

This was not the only ground-breaking innovation that can be traced back to Faraday. Ørsted's proof that electricity can cause magnetism led him to consider whether electricity could not also be produced from magnetism. In 1831, Faraday wound a metal wire around half of an iron ring and connected it to an electric circuit that he could switch on and off by means of a switch. A second metal wire was wound around the remaining part of the iron ring without touching the first wire. He connected a current-measuring device to this second wire. Faraday hypothesised that the current flowing through the first wire would magnetise the iron ring and that this magnetism must in turn produce an electric current in the second wire. Sure enough, the ammeter tripped on the second wire. To his surprise, however, the deflection went down again immediately after the switching process in the first circuit, although current continued to flow. Even more astonishing was that a deflection occurred again when Faraday switched off the current in the first circuit, and this also immediately decreased. Faraday had discovered the principle of "electromagnetic induction": only a magnetic field that changes or moves in time causes an electric current.

In another experimental arrangement, he therefore let a coiled copper wire rotate in the field of a magnet. In this way, he created a magnetic field that constantly changed relative to the copper wire and induced a steady (alternating) current. Faraday had invented the dynamo, and with it a way to convert mechanical energy into electromagnetic energy. Faraday could not have foreseen the impact his discovery would have. But it was clear to him that it was significant. When a minister asked him what could be done with his discovery, he replied that he did not know, but was convinced that one day taxes could be levied on it.

For Faraday, all forces in nature—electricity, magnetism, heat, light, gravitation, etc.—were different manifestations of a single fundamental force. His deep belief in the unity of nature and the conviction that all forces can be transformed into one another guided him in all his physical experiments. This speculative and intuitive approach to physics was typical of the German and English philosophers and naturalists. In France, on the other hand, a rational, mathematically based form of physics prevailed; here the romantic philosophy of nature of Goethe, Faraday, and Ørsted was referred to as a "rêverie allemande", i.e., a German reverie.

The Last Riddle

The scientific world, which relied on pure reason, experiments, and precise calculations, reacted to Faraday's conception of electromagnetic fields with great scepticism, because his field theory lacked just the thing that distinguished Newtonian physics: a comprehensive mathematical foundation and integration into an overall theoretical system. This would only change with one of the most ingenious individual achievements in the history of physics: shortly before Faraday's death, the Scotsman James Clerk Maxwell succeeded with his "electromagnetic field theory" in describing all known phenomena of electromagnetism with mathematical precision, thereby integrating the phenomena of electricity and magnetism, long considered mystical, into the rational explanation of the world.

At the age of 15, Maxwell had already impressed the scientific community with brilliant mathematical explanations. In 1855, when he published his field theory in the form of a simplified mathematical model, he was only 24 years old. In it, he reduced all contemporary knowledge of electricity and magnetism to a single linked set of differential equations consisting of 20 equations and 20 variables. He made his final theory available to the public in the form of four equations in 1873, in his work "Treatise on Electricity and Magnetism". Two of Maxwell's equations describe static electric and magnetic fields, and the other two the dynamic interactions between them:

- Equation 1: Electric charges at rest cause electric fields.

- Equation 2: Magnetic field lines are always closed.
- Equation 3: Moving electric charge (i.e., electric current) causes a magnetic field.
- Equation 4: Temporal variations of a magnetic field generate an electric field.

The proof that Faraday had been right with his idea of electromagnetic fields could have been a great triumph for romantic natural philosophy. But at the same time it brought about its end, because the strict mathematisation of Faraday's field theory left no room for any vague philosophical thoughts.

Small Problems and Big Contradictions

With Maxwell's field theory, physics had explained all known phenomena and made them calculable. Anything further—for example, proving that all forces can actually be transformed into one another—would have been nothing more than hard work with an expected outcome. It was a new high point of classical physics—and at the same time its last. For Maxwell's equations were like a Trojan horse that carried its demise in the form of two problems:

- The aether problem
- The problem of the speed of light.

The aether problem. To his great surprise, Maxwell realised that the solutions of his equations not only described known phenomena, but also made predictions of entirely new phenomena. The most significant of these was that electric and magnetic fields can detach themselves from their source and propagate freely in space. Their movement corresponds to a wave motion that results from an up and down variation of the electric and magnetic fields, analogous to the crests and troughs of water waves. An electric field oscillates in a plane, and a magnetic field oscillates in the same plane but rotated by 90 degrees and shifted by a trough of the wave.

According to the ideas of classical physics, these electromagnetic waves had to obey the laws of mechanics. Water waves need water to propagate, sound waves need air, for example, and mechanical waves travel along a curved rope, for example. But in which medium do electromagnetic waves oscillate? This all-pervading, invisible substance had to be the aether already mentioned at the beginning of this chapter. Aristotle had already "invented" it. For Newton and Faraday, too, it was inconceivable that there could be an absolute void in nature. But no matter how hard the natural scientists tried, none of them had ever been able to prove or even measure any properties of the aether in their experiments. This problem could have been postponed with the argument that one day a researcher would come up with a brilliant idea about how to identify the aether. The only fatal thing would have been to prove that the aether did *not* exist. But with the second problem, it was the other way round: it was solved—and everything collapsed.

The problem of the speed of light. According to Maxwell's calculations, electromagnetic waves had to propagate at a constant speed. His equations even provided the value of this speed: in a vacuum, it is around 300,000 km/s. But in 1849, the French physicist Hippolyte Fizeau had succeeded in measuring the speed of light quite accurately with the help of a mirror and rotating gears, and it was also about 300,000 km/s. This could not be a coincidence! Maxwell ventured the hypothesis that light consists of electromagnetic waves.

This speed of light caused physicists great headaches. Since Galileo, in connection with the principle of inertia, the irrefutable rule had applied that all natural events behave independently of whether a system is in a state of rest or uniform motion. As a logical consequence, velocities add up: a stone thrown at a speed of 30 km/h from a train travelling at 100 km/h in the direction of travel has a speed of 130 km/h from the point of view of an observer standing by the railway tracks. This way of calculating also agrees with Newton's laws. But Maxwell's equations do not allow such an addition; according to them, the value of the speed of light must remain constant at 300,000 km/s, regardless of whether the observed system is in motion or not. What appeared to be a single, tiny issue was the touchstone for whether Maxwell's electromagnetic theory, which offered such a wonderful complement to Newton's laws, was universally valid. If it proved to be correct, however, Newton's world formula would lose its validity. Either way, physicists would lose a lot.

Maxwell and his contemporaries did not yet dare to conclude that the speed of light could actually be constant and independent of the state of motion of the source. This would have put them on course for a direct confrontation with Newton. They saved themselves with an auxiliary hypothesis in which the aether was supposed to solve the problem that had arisen:

- With the aether, electromagnetic fields and waves would have a medium of propagation, just as the Newtonian world envisaged.

- Aether should also represent the absolute system of rest, every uniform motion would be described in relation to it.
- In the aether, Maxwell's equations should apply exactly, i.e., the speed of light should assume its theoretical value of 300,000 km/s. In all other uniformly moving systems, it should deviate from this.

Just as rubbish left in the cellar eventually sends its smells through the house, the problems which physicists had wanted to leave behind with the unproven aether hypothesis would soon catch up with them. To their great dismay, in 1887, eight years after Maxwell's death, an experiment conducted by two Americans in Chicago showed that Maxwell's equations were correct and Newton's laws were not. The physicist Albert Michelson and the chemist Edward Morley managed to determine the difference between the speed of light in the direction of the Earth's motion and that in the opposite direction. On its orbit around the Sun, the Earth has a speed of almost 30 kilometres per second. The speed of light should therefore have been faster by this amount in one measurement and slower by this amount in the other. But the experiment proved beyond doubt that the speed of light was identical in both directions, even though the direction of the light was once with and once against the Earth's motion. Something fundamental was wrong with Newtonian physics. All certainty was gone in one fell swoop. The ground had been pulled out from under the physicists' feet, and there was no alternative explanation in sight.

2

The Battle for the Atom—From Boltzmann to Einstein—How Chance Broke into the Well-Ordered World of Physics

At the very time that Michelson and Morley experimentally demonstrated that the speed of light was constant, another certainty of physics collapsed. Newton's laws, and later Maxwell's electrodynamics, had presented the world as a more predictable place. Physicists agreed that only a few last little problems needed to be solved before they could understand nature in all its details. There was even the idea that one only had to know all the "cogs" of the world precisely in order to be able to predict every event exactly until the end of time. The great French mathematician Pierre-Simon Laplace had already put this idea into words in 1814 in his book "Philosophical Experiment on Probabilities":

> We can consider the present state of the universe as the effect of its past and the cause of its future. An intelligence which at a given moment knows all the forces which set nature in motion and all the positions of all the elements of which nature is composed, if this intellect were also great enough to subject these data to analysis, it would sum up in a single formula the movements of the largest bodies in the universe and those of the smallest atom; for such an intellect, nothing would be uncertain and the future as well as the past would be present before its eyes.

Such an intelligence was later called Laplace's demon. That all-encompassing knowledge was possible in principle was not doubted. The new research field of thermodynamics would soon shake this certainty.

L. Jaeger, *The Stumbling Progress of 20th Century Science*, https://doi.org/10.1007/978-3-031-09618-1_2

Locomotives and Hot Water Bottles

The increasing demand for efficient steam engines meant that it became more and more important to understand the process of heat exchange. It was known from experience that the more heat was supplied to a machine, the faster it would operate. By the middle of the nineteenth century, it became clear that this was a transformation between different forms of energy: the heat used to fire the machine was transformed into mechanical energy in the form of a pounding piston, minus a certain amount of heat loss. Now it was not a big step to realising that other forms of energy could also be transformed into each other. In 1842, the Heidelberg physician Julius Robert Mayer published one of the most important theorems in physics:

> My contention is that falling force, motion, heat, light, electricity, and chemical difference of ponderables are one and the same object in different manifestations.

This is the First Law of Thermodynamics: energies are only ever converted from one form to another; they can never disappear or arise from nothing. With this "law of conservation of energy", physicists were continuing along the path first taken by Newton, who had already shown for mechanics that no motion can come about from nothing. Therefore, the deeply rooted belief in Newton's laws was not yet in danger.

However, physicists and engineers quickly discovered a special feature of thermodynamics. While in "Newton's world" all processes can move back and forth—a weight on a (frictionless) pulley can be raised and lowered any number of times—energy transformations in thermodynamic processes are only possible to a limited extent: the heat of a body can only be converted into mechanical energy if its temperature is *higher* than that of its surroundings. For example, a locomotive gains its energy from the heat of burning coal. However, the First Law of Thermodynamics only requires that, on balance, the sum of energies must remain the same; it does not prohibit a relatively cold body from becoming even colder by transferring energy to a warmer body. So there had to be a second fundamental law of heat that would prevent a locomotive from setting itself in motion by diverting energy for itself from cold ambient air. In 1850, the German physicist Rudolf Clausius formulated a theorem that later became what we know today as the "Second Law of Thermodynamics".

> There is no change of state whose only result is the transfer of heat from a body of lower to a body of higher temperature.

This means that the conversion of heat into mechanical energy will only work until the energy-transferring body has cooled down to the temperature of the energy-receiving system. As soon as both bodies have the same temperature—for example, when the hot water bottle has cooled down to the body temperature of the person using it to keep warm—the heat transfer will come to a standstill. In the system, heat exchange will only take place again when new energy is supplied from outside, for example, by refilling the hot water bottle with hot water.

To describe this limitation mathematically, Clausius introduced the term "transformation potential" into physics, also called "entropy". Through trial and error, he found an equation in which entropy is defined as the ratio of the amount of heat transferred to the difference in temperatures at the beginning and end points of the heat transfer. With the help of this formula, the events in a steam engine could finally be calculated. The Second Law of Thermodynamics could now be formulated as follows:

> In a process of heat transformation between two systems, entropy can only increase, never decrease.

When the entropy has reached a maximum value, the heat transfer ends. Analogously, this means that the temperature differences in a closed system can only become smaller, not larger.

It was known that the mathematical calculation of entropy worked, but it was not yet known *why* this was so. A fundamental theory of the nature of heat did not yet exist. Which processes in nature led to the entropy law remained a mystery.

Chance Enters the Stage of Physics

By the end of the nineteenth century, the prevailing view in physics was that, ultimately, all physical phenomena in the microcosm could be traced back to electromagnetic events. According to the wave nature of electromagnetism and the continuity of the solutions of Maxwell's equations, all matter should be of a continuous nature, i.e., divisible to infinity. Only a few physicists held onto the idea that matter consists of indivisible and massive spheres. They imagined that these atoms would swirl around in a gas like tiny billiard balls, and in numbers that would defy imagination. They asked themselves:

- Could heat transport be related to the mechanical motions of the smallest particles?

- How could the processes in the micro-world of gases be described mathematically so that one could calculate the quantities measured in the macro-world, such as pressure, volume, and temperature?
- How should one think of entropy in terms of atoms?

It was clearly impossible to take into account the behaviour of each individual atomic sphere according to Newton's laws. But then an unexpected breakthrough was made, inspired by James Clerk Maxwell., the father of electrodynamics: astonishingly, it was discovered that one does not need to know the motion of each individual particle at all, but only the mean values of the motions of all the particles. It is like a huge crowd of people moving out of the stadium at the end of a football match. Although each individual leaves at very different speeds—some push their way to the front, others wait for acquaintances at some agreed meeting point—statistical equations and only a few parameters such as average speed, direction, and variance of the motions can be used to predict very precisely how long it will take for the stadium to empty.

The fact that statistics could become the mathematical tool of heat theory shook the firmly established edifice of Newton's classical physics. This was not so much because once again, through trial and error, a way had been found to describe nature exactly without having any idea *why* it worked. Physicists had almost got used to that. The introduction of statistics was revolutionary because one had to assume that the smallest particles do not move predictably according to Newton's laws, but purely by chance. Doing exact physics with probability considerations seemed absurd to physicists for whom Newtonian determinism was still sacrosanct. In addition, the use of statistics necessarily presupposed the existence of atoms. Only a few nineteenth century physicists were ready to accept this idea. Among the most vehement opponents of atomic theory were the physicist Ernst Mach in Austria, the young physicist Max Planck in Germany, and the great mathematician Henri Poincaré in France. Their aversion had mainly philosophical reasons:

- An atom must occupy a certain region of space. However, because man's conception of space is continuous, we can also *imagine* a half, a quarter, and a thousandth of an atom in a thought experiment. This means that whole atoms are by no means (mentally) indivisible. On the other hand, matter must be composed of the smallest particles because this is the only way to ensure its firm hold. Without these massive basic building blocks, all matter would have to dissolve like water. Atoms must therefore exist and at the same time they cannot exist at all. This contradiction was already familiar to the Greeks and had been reaffirmed by Immanuel Kant only about a hundred years earlier.

- Kant's philosophy states *why* this contradiction exists. For him, it was unquestionable that we can never find answers outside our own limits of experience. As soon as we leave the world that is directly visible and tangible to us, we necessarily encounter inner contradictions. This view was held in particular by Ernst Mach. According to his conviction, real knowledge could only be derived from direct experience or from experiments. When he was asked about the existence of atoms, his standard answer in his typical Viennese dialect was: "Ham S' ans g'sehen?"—Have you seen one? (After the development of the scanning tunnelling microscope in 1981, this question could be answered with a resounding "Yes").

Do atoms exist or not? In physics, this discussion was fought tooth and nail well into the twentieth century. Remarkably, there was no such intellectual dispute in chemistry. Chemists were much more pragmatic and hardly bothered with such philosophical questions. While physicists were still at loggerheads, the existence of atoms had been accepted in chemistry as a self-evident and indispensable foundation for the subject from the beginning of the nineteenth century, i.e., for almost a hundred years.

A Gravestone at the Vienna Central Cemetery

Statistical calculations helped engineers to build better machines in practice. But the theoretical foundation that explained why this type of calculation was applicable was still missing. It was not established until the Austrian physicist Ludwig Boltzmann succeeded in creating the theoretical mathematical foundation of thermodynamics. At the age of only 22, he recognised the crucial importance of entropy, introduced by Clausius, and the way it relates to the statistical observation of processes at the atomic level. In a brilliant move, Boltzmann defined entropy in 1866 in his work "On the Mechanical Meaning of the Second Law of Thermodynamics" as the *number of possible states in which the particles contained in a system can arrange themselves in order to produce, in sum, the macro-state perceptible by man*. Boltzmann's decisive formula for the entropy S is, in simplified form,

$$S = k \ln \Omega$$

Ω represents the number of possible microstates of the system in a given macrostate, and the proportionality factor k is the Boltzmann constant $(1.381 \cdot 10^{-23}$ J/K). The entropy S is therefore proportional to the logarithm of the number of possible microstates in the system. The greater the entropy, the more microstates are possible and the greater the disorder within the system. The Second Law of Thermodynamics was directly derivable from the fact that disordered states are most likely in a world of mechanically colliding particles. For a system to transform itself from a disordered to an ordered state without external influence is just as improbable as toothpaste squeezed out of a tube finding its way back into the container. Boltzmann turned thermodynamics into "statistical mechanics". In the new reading of the Second Law of Thermodynamics, the number of possible microstates of a system—and thus the disorder—increases steadily until it has assumed the greatest possible value.

With Boltzmann's trick, thermodynamics could be explained on the basis of individually moving, randomly distributed atomic particles undergoing collisions and being subject to Newton's laws.

- If they move quickly and wildly in a hot gas or a hot liquid, or if they oscillate strongly back and forth in a solid body, the disorder and thus the entropy is high.
- At low temperatures, on the other hand, the particles of a gas or a liquid move only slowly, or vibrate little in a solid body. Hardly any energy can then be extracted from their low level of motion and the entropy of the system is low.

Entropy S: The more heat a system can potentially transfer to another system, the greater its entropy. Heat means that the particles move quickly and are disordered. Thus entropy is also a measure of the disorder in a system and the number of states it can be in.

Boltzmann's explanation of entropy is taught in schools today, but during his lifetime he had to endure fierce attacks from physicists who could not or would not accept his thermodynamics. For with chance a key ingredient, physics was no longer strictly deterministic.

Depressed by decades of hostility from his opponents, but also by his failing eyesight and—as he perceived it—his flagging mental abilities, he took his own life in 1906, at the age of 62. The equation he found, comparable in rank to Newton's laws of force and Maxwell's equations of electrodynamics, is carved into his tombstone in Vienna. Only a few years after his death, the existence of atoms was demonstrated experimentally.

Even though many scientists still distrusted statistical mechanics around 1900, with the trio of Newton, Maxwell, and Boltzmann leading the way, classical physics had long been complete. Everything seemed to mesh together beautifully and, for those who could come to terms with the existence of chance in physics, the world was all right again for a brief respite. It was the American physicist Josiah Willard Gibbs who discovered a fundamental problem within Boltzmann's definition of entropy. The following experimental setup shows what the conflict is.

> **The builders of classical physics:**
> - **Newton** laid the foundation with his mechanics
> - **Maxwell** integrated electrodynamics into classical physics
> - And thanks to **Boltzmann,** thermodynamics was now also part of Newton's world, although only by allowing chance to become part of classical physics.

Two adjacent containers of the same size contain the same gas at the same temperature and pressure, each of them has the same entropy value S, so overall the two-container system has a value of 2S.

- Now a door is opened between the containers so that the gas particles can mix. According to Boltzmann's equation, the entropy of the gas in the two-container system is now no longer 2S, but increases abruptly. This is because particles from the different containers can now change places with each other. Each exchange enables a new microstate of the mixture, while the macrostate remains the same. Despite this considerable mathematical increase in the number of possible microstates and thus the mathematical increase in entropy, we see that the pressure, temperature, and volume remain the same in the macro world.
- After closing the door, the entropy would be reduced again to 2S within the framework of Boltzmann's calculations, but there cannot be a decrease in entropy according to the Second Law of Thermodynamics.

There was no doubt: the explanation Boltzmann had offered was very well suited for calculations and could predict the results of experiments, but it failed the reality check. Once again, physicists had put together a 5000-piece puzzle, so to speak, and the picture was perfect, but there were other pieces of the puzzle lying around on the table for which there was no room in the picture. In order to hold on to the finished and coherent picture, they would have to sweep the excess pieces under the table. They had already had to go through the same thing when it had been shown that the speed of light was not additive. For all those who felt committed to Newton, it was another devastating setback.

Ludwig Boltzmann, who wanted to reconcile thermodynamics with Newton's world, was the last great builder of classical physics. At the same time, he was the forerunner of a completely new way of interpreting the world, the outlines of which were just beginning to emerge from the fog at that time, and which would later lead to quantum theory. For Boltzmann's theory and Gibb's problem signalled a chasm between the old and the new physics, becoming so deep that the coming generation of physicists *had no choice* but to dare to jump to the other side:

- Boltzmann used the term "quantum" as early as 1877 in his description of the statistical distribution of the energies of gas molecules, specifically as a mathematical term for the description of energy ("energy quanta"). But his quanta were a purely calculated quantity without any physical meaning. Unfortunately, after 1900, he did not publish anything more on the quantum problem; we do not know whether he was able to do anything with the idea of discrete energy units. Planck's quantum relation $E = h \cdot f$, which will be discussed below, was formulated during his lifetime.
- Max Planck, known today as the "father of quantum physics", also initially did not consider quanta to be real, according to his own admission. Planck's quantum formula was inspired by Boltzmann's mathematical trick, but unlike Boltzmann, he could no longer allow his quanta to tend towards zero despite his most determined efforts.[1]

Boltzmann's student Lise Meitner formulated its role in physics thus:

With his thermodynamic research and the introduction of statistical methods, he contributed significantly to the transition from classical to modern microphysics.

It is tragic that Boltzmann took his own life precisely at this decisive interface with modern physics. Had he been able to continue his research, he is probably the one who would have gone down in history as the father of quantum theory.

---

[1] See also: U. Hoyer, *Ludwig Boltzmann und das Grundlagenproblem der Quantentheorie*, Zeitschrift für allgemeine Wissenschaftstheorie; Vol. 15, No. 2, pp. 201–210 (1984).

Planck's Act of Desperation

The first physicist to leave Newtonian physics behind—involuntarily but consciously—was Max Planck. His field of research was the radiant energy emitted by solid bodies. It had been known for millennia that the colour spectrum of incandescent bodies shifts from red to yellow and blue towards white with increasing supply of heat. But why is this so? Planck also wanted to know:

- How do bodies absorb and release energy?
- How does energy depend on the temperature of the body?
- What is the exact frequency of the emitted light?

His colleague Gustav Kirchhoff, also working at Berlin University, had shown in 1859 that the frequency distribution of radiation emitted by heated bodies is independent of the material used and depends solely on the temperature of the body. He had also introduced the concept of a "black body". This is an object that absorbs all radiation hitting it. A piece of coal comes quite close to this property because it does not reflect any radiation in the frequency range of visible light, i.e., it is black. If the black body absorbs all electromagnetic radiation, it should also be able to emit all radiation frequencies due to the possibility of reaching thermodynamic equilibrium.

Planck wanted to use the mechanical statistics that Boltzmann had developed for gases to derive the macroscopically measurable properties of a black body from its micro level structures. But because he was strongly influenced by Ernst Mach, Planck kept his distance from Boltzmann's atomic ideas and assumed a continuous, "atom-free" distribution of energies in the black body. Despite intensive efforts, he did not succeed in establishing a formula for the radiation energy of ideal black bodies under this assumption, at least not one whose results agreed with measurements. Using the common continuous wave model of energy, it was mainly high frequencies that should have been observed in the radiation emanating from such a body, while experiment showed that the high and low frequencies of the radiation energy were relatively evenly distributed.

After long and fruitless efforts, Planck, in an "act of desperation", as he put it himself, made an additional assumption: he postulated that energy is not radiated continuously, but in packaged portions, so-called quanta (from the Latin *quantum*, meaning "so much"). This assumption fundamentally contradicted classical physics, in which all electromagnetic radiation was of a continuous nature. There were also philosophical objections to the idea of quanta. Over 300 years earlier, Leibniz had formulated: "Nature does not make leaps." But assuming that the energy of these quanta depended directly on the frequency of the electromagnetic radiation, Planck succeeded in deriving a formula that exactly reflected the experimental observations:

$$E = h \cdot f$$

This connection between energy and *frequency* was something completely new. Until then, it had been assumed that energy was determined by *amplitude*, i.e., the intensity of the electromagnetic waves. Planck's formula even offered an explanation as to why high frequencies occur less frequently in experiments than originally expected: they require more energy.

Planck also succeeded in deriving the numerical value of the constant k in Boltzmann's law in his model. He called it the "Boltzmann constant". In fact, Boltzmann had already derived the distribution of energy among the atoms of a gas in 1877, considering finite energy intervals, thus already introducing a kind of quantisation of energy. Although he regarded this as a mere mathematical trick, Planck was undoubtedly familiar with this procedure. Was he perhaps even guided by it?

Unfortunately, the correspondence between Boltzmann and Planck was lost during the Second World War. But Planck wrote to Boltzmann, whose atomic theory and statistical physics he had long rejected, about his atomistic successes in radiation theory. The latter showed Planck his interest and accepted his approach. In his later writings, Planck again fully emphasised his admiration for Boltzmann.

On 14 December 1900, Planck presented his formula and the quantum hypothesis on which it was based to the German Physical Society. Today, this day is considered to mark the birth of quantum theory—and at the same time the beginning of the final farewell to Newton's physics. However, there was no talk of a revolution at first. Planck himself regarded his hypothesis only as an auxiliary construction without deeper relevance, and something he hoped later to get rid of. Accordingly, he gave his quantum of action the mathematical symbol h for "auxiliary quantity". He desperately tried to derive his radiation formula without the unwelcome quanta. But his auxiliary quantity was never to disappear from physics and has retained its name to this day.

Planck's formula gave physicists quite a headache, because energy in this context is a property of particles, frequency a property of waves, whereas here both were combined in *one* formula. So, what were quanta? Particles or waves? Planck's quantum hypothesis was left to one side for a few years. It was all too crazy, too wacky, no one expected it to have much future!

Einstein's Solution and Other Fundamental Contradictions

The first interpretation that viewed Planck's quantum of action, not as a pragmatic mathematical trick, but as a physical reality, was ventured by Albert Einstein. Unlike most of his contemporaries, he noticed early on and with astonishing clarity the creaking in the woodwork of the three then known subfields of classical physics—mechanics, electrodynamics, and thermodynamics. All the disturbing noises could be traced back to the same source: events in the microcosm seemed to elude the laws of classical physics.

In 1905, Einstein was an unknown 25-year-old physicist who was working full-time as a clerk at the patent office in Bern, as a "Tintenscheißer" (ink shitter), as he described himself. Five years after Planck had brought his embarrassing solution into the world, Einstein went a decisive step further: in his paper „Über einen die Erzeugung und Verwandlung des Lichts betreffenden heuristischen Gesichtspunkt", "On a heuristic point of view concerning the production and transformation of light", submitted in March 1905, he claimed that:

- *All* electromagnetic radiation occurs in quanta, including light.
- Energy packets behave like spatially localised particles.

Tremendous confusion arose, because in classical physics it had been proven in numerous experiments that light is definitely a wave. But there was also a key witness for Einstein's hypothesis: while with waves it is their amplitude that matters, with particles their frequency plays the main role. And there was more and more evidence for the importance of frequency:

- In 1887, Heinrich Hertz discovered the photoelectric effect. He irradiated metals with electromagnetic radiation and measured the number of particles knocked out of the metal with a particle detector (later it turned out that they were electrons). In a side note to his experiments, he observed that, below a certain frequency of the incident light, no more electrons emerge, no matter how high the radiation intensity. However, if light is a wave, the energy of the ejected electrons should not depend on the frequency, but exclusively on the amplitude (intensity) of the incident radiation. Hertz did not pursue this inconsistency.

- Since 1900, thanks to Planck's formula $E = h \cdot f$, the relationship between energy and frequency of radiation was already on the table, it just hadn't received any attention yet. Planck himself was highly dissatisfied with this result. Only Einstein took the formula seriously.
- In 1902, the Austro-Hungarian physicist Philipp Lenard had proven that the energy of the emitted electrons in the photoelectric effect is also independent of the intensity of the electromagnetic radiation. This clearly spoke against a wave nature for electromagnetic radiation.

Einstein's hypothesis that energy quanta appear as particles would also eliminate a puzzle that had been bothering physicists since Maxwell: unlike classical waves, they do not need a medium in which to move. With quanta, he could finally get rid of the idea that there is a substance called aether.

Einstein explained his hypothesis like this: a light quantum penetrating the metal gives up its energy, which is directly proportional to the frequency of the radiation according to Planck's formula $E = h \cdot f$, either completely or partially to an electron in this metal. Thus, the now free electron has the energy $E = h \cdot f - P$, where P is the energy that was needed to release the electron from the atomic bond. For photons with an energy below P, no electron escapes. With this simple model, Einstein was able to describe the experimental observations of Hertz and Lenard exactly.

Proof of Einstein's hypothesis was to be provided by measuring the exact amount of energy of the escaping electrons. Would the measurement result of the experiment agree with the calculations? For nine years, Einstein's equation could not be confirmed experimentally; it was possible to measure the energy of the incident radiation precisely enough, but not that of the escaping electrons. It was not until 1914 that the law predicted by Einstein was measured with sufficient precision by the US physicist Robert Millikan. Einstein was awarded the Nobel Prize in Physics in 1921 for his prediction and explanation of the photoelectric effect. His light particle hypothesis anchored Planck's quanta in physics. Now Planck's quantum of action was no longer an auxiliary construction, but it affected the whole of classical physics. With it, the processes in the macro world could be wonderfully explained and calculated, but as soon as physicists left the realm of directly observable phenomena and turned to the micro world of atoms, classical physics had proven to be useless.

Einstein's Second Master Stroke

Under the title „Über die von der molekularkinetischen Theorie der Wärme geforderte Bewegung von in ruhenden Flüssigkeiten suspendierten Teilchen" ("On the motion of small particles suspended in stationary liquids required by the molecular-kinetic theory of heat"), Einstein submitted another ground-breaking paper in May 1905, which became the basis of his dissertation at the University of Zurich. It dealt with the problems of Ludwig Boltzmann's statistical mechanics and was the result of his first scientific efforts from 1902 to 1904. While statistical mechanics, which derives the macroscopic properties of gases from the average motion of *countless particles*, was still controversial, Einstein had already gone one step further. He asked himself how systems behave when they consist of so few particles that the fluctuations between them do *not* average out and therefore statistical mechanics is *not* applicable. To do this, he needed to know what the properties of particles are and how they move in concrete terms. Einstein's earliest scientific considerations thus immediately addressed the most important problem of physics of the early twentieth century: he asked about the structure of matter.

Even though the evidence for the existence of atoms had increased, at the beginning of the twentieth century neither physicists nor chemists could say anything concrete about their nature, especially not about their dimensions and weights. Einstein began his reflections by investigating a hitherto neglected phenomenon discovered by the Scottish botanist Robert Brown in 1827. The latter had noticed during microscope observation of plant pollen dispersed in a liquid that the pollen grains were in constant, uncoordinated motion. The warmer it was, the more the pollen jiggled around. Einstein assumed that this "Brownian motion" was the result of fluctuations of the smallest particles in the liquid. Based on this assumption, he derived a mathematical formula that related Brownian motion and the size of the particles to measurable quantities such as heat conduction in liquids and diffusion in gases. Four years later, the French physicist Jean Baptiste Perrin confirmed Einstein's formula in experiment. This was the first concrete proof of the existence of atoms and at the same time gave a method for determining their size.

But how could Einstein demonstrate the motion of individual particles? His speculation with pollen was not enough to convince his contemporaries. One year before his suicide Boltzmann was of the opinion that they were too small ever to be understood. Einstein shared this view and hoped to find a system in which the behaviour of a single particle could nevertheless be understood. It is one of the greatest strokes of luck in the history of physics that he came across Planck's "black bodies" in his search.

The Third Revolution from the Bern Patent Office

In June of 1905, a year so crucial for physics, Einstein submitted a third paper—it too appeared in the journal "Annalen der Physik" edited by Max Planck. This time it was about the properties of electromagnetic fields and their effects on the nature of time and space. Under the name "Theory of Relativity", it immediately attracted great attention among theoretical physicists.

The starting point for Einstein's considerations was the invariance of the speed of light. The situation after the Michelson–Morley experiment was that the classical law of velocity addition does not apply to light. But what is speed anyway? In everyday life, and thus also in classical physics, it is the ratio of spatial to temporal change, for example, when we say "100 km/h". Space and time are regarded as two separate spheres. This separation was the basis of Newtonian physics. For Newton, as for Galileo before him, space was like an "external container" completely independent of time, in which all motion takes place. Some of Newton's contemporaries, especially his rival Gottfried Wilhelm Leibniz, were highly sceptical about this idea of an absolute space and time. The idea that time and space might be interdependent had thus already been speculated about two hundred years earlier. Einstein, however, was the first person to think radically about how space and time actually behave and how we have to change our ideas if we want to understand space and time realistically. For if (light) velocities cannot be added up, it is not only the idea of velocity that must be revised, but also that of space and time, in such a way that space and time components become directly connected to each other to form "space–time". Einstein stated that four direct consequences arise from the concept of space–time:

- A moving body appears foreshortened to an observer at rest relative to it.
- Time is not the same for all systems, but depends on the spatial motion of the system in which it is measured. Time in moving systems runs more slowly than in systems at rest relative to it. This relativity of time and space gave Einstein's theory its name.
- No body moves faster than light. The speed of light is an upper limit for all physically possible speeds.

- The mass of a particle increases with increasing speed. This realisation eventually led to Einstein's famous formula $E = mc^2$. The equivalence of mass and energy in the theory of relativity went hand in hand with Planck's quantum hypothesis. So it was by no means an accident that Einstein was working on these two topics at the same time.

Einstein chose the title "On the Electrodynamics of Moving Bodies" for his publication on the theory of relativity, because the starting point of his new theory was not railway stations, travellers, and trains, but the properties of electromagnetic fields in moving systems. To approach this, Einstein started from a simple idea. Through Faraday's experiments and Maxwell's equations, he knew that:

- A *moving* electron generates a magnetic field, whereas an electron at rest generates a static electric field.
- The speed of light is constant.

But if velocities cannot be reliably added up—and the constancy of the speed of light under all circumstances means precisely that—then there can be no absolute motion. And thus there can be no generally valid difference between an electron in motion and an electron at rest. The idea of absolute space and absolute time therefore had to be replaced by something completely new. While in Newton's world the electric field ("resting" electron) and the magnetic field ("moving" electron) are completely different phenomena, in the real world beyond the limitations of classical physics, each had to be transformable into the other. This is exactly what Einstein proved: magnetic and electric fields could be transformed mathematically into each other with his new rules for space–time transformations.

Wanderer Between Two Worlds

Thanks to Planck's and Einstein's work, it became apparent that the conditions in the microcosm are quite different from those in the macrocosm, which is the world we are familiar with and can experience. Each approached the question of the nature of the microcosm from a different angle: particle or wave?

- Planck's formula $E = h \cdot f$ established a connection between energy occurring in the form of quantum particles and a wave frequency.

- Einstein's formula $E = m \cdot c^2$ links the mass of a particle with the inherent energy it contains, which is a typical wave property.

While the vast majority of physicists, including Planck, were still completely anchored in the thinking of classical physics until the 1920s, Einstein became a wanderer at an early age, travelling between the world of classical physics, in which he had grown up and been educated, and the world of a new physics, which—as he was the very first to suspect—had dramatically different properties and laws. He consistently separated the macro- and micro-worlds:

- As long as we are dealing with relatively large spatial structures, the quantum nature of light is not noticeable. Then light manifests itself as a wave. In a macroscopic system, one does not have to worry about the properties of individual particles. The macroscopic properties of the system can be determined statistically from the average properties of countless particles—this applies to gas atoms (thermodynamics/Boltzmann) just as it applies to electromagnetic radiation (electromagnetism/Einstein). "Observations refer to temporal averages and not to instantaneous values", Einstein said in 1905. Thus, the experimental phenomena so typical for the wave nature of light, such as interference patterns or diffraction, which are so successfully described by classical physics, arise in the macro world.
- In the microworld, on the other hand, the system consists of so few particles that their properties cannot be described with mean statistical values. Here, the non-continuous nature of the quantum world must be taken into account and energy packets must be calculated. The quanta correspond to photons in light and atoms in matter. The attempt to calculate events with "continuous space functions" as in the macro world leads to unsolvable difficulties in the micro world. This is because at the atomic level, radiation emission and absorption are events localised at a point, so mathematical space functions are not possible.

For Einstein, the established wave theory of light in the macro world existed in parallel with the particle theory in the micro world. For him, they were two sides of the same coin. But even with this approach, an insoluble question arises: How must one imagine a spatially localised quantum of light? Does it have a spatial extension?

- Without spatial expansion, a quantum of action would have to have an infinitely high energy density, an idea that is difficult to convey to a physicist. The same applies to electrons. If an electron occupies no space, it must have an infinitely high charge density and thus produce infinitely large forces.
- If these entities had a spatial extension, there would be spatial continuity and particles or quanta would have to be divisible. It would also be possible to imagine a spatial function *within* the light quantum or electron. Einstein rejected this view.

Philosophers like Democritus and Kant had already encountered this problem in their purely mental apprehension of atoms. The 2500-year-old mental dilemma had only slipped one (spatial) level deeper at the beginning of the twentieth century: it was now no longer a question of whether there are indivisible atoms, but whether electrons and photons are indivisible.

Einstein could not provide a satisfactory explanation of the world. He had left the shore of classical physics, but he remained far from the opposite shore of a completely new physics. The following anecdote provides an apt picture of Einstein's position in the history of physics: when he was once asked whether he was standing on Newton's shoulders, he replied: "No, on Maxwell's shoulders." That something was wrong with Newton's worldview was obvious. But the theoretical foundation of relativity was Maxwell's classical equations. Einstein was unable to break away from this anchor.

The Contours of the Atom—Further Contradictions

Physicists' theoretical considerations had plunged them into a confusion that was hard to bear. Nothing fitted together and everything called into question the securely believed foundations of classical physics. The more knowledge was accumulated, the more the picture of the world became blurred. It was exasperating!

While the most brilliant minds in theoretical physics manoeuvred themselves into a dead end from which they could no longer find their way out, experimental physicists took the lead in trying to understand the world. Step by step, they approached the true nature of atoms:

- The English researcher Joseph John Thomson was concerned with so-called "cathode radiation". It was discovered that, when an electric field is applied to a metal, radiation emerges from it which, unlike light, can be deflected by magnetic fields. Thomson concluded that these particles, which he called *corpuscles,* must be electrically charged. In 1891, the Irish physicist George Johnstone Stoney proposed the name "electrons" for them. Thomson assumed that the electrons were components of the atoms—a completely new and very radical idea, because this would mean that the atom was no longer indivisible.

- In 1895, the German physicist Wilhelm Conrad Röntgen discovered a completely new form of radiation during experiments with cathode radiation, which he called "X-rays". They even passed through the human body, and today they are called X-rays in his honour.
- Inspired by Röntgen's work, the French physicist Henri Becquerel described the phenomenon of radioactivity shortly afterwards.
- His compatriots, the couple Marie and Pierre Curie, painstakingly proved that radioactive radiation is associated with the transformation of elements.

Further evidence was gathered that the atom consists of different building blocks. Another breakthrough was made in one of the most famous experiments in the history of physics. The New Zealand physicist and student of Thomson, Ernest Rutherford, bombarded a thin gold foil with positively charged helium nuclei and measured the resulting radiation with a photographic plate built around the gold foil. This revealed something strange: some radioactive particles were strongly deflected by the thin gold foil, in some cases even reflected by almost 180°. This result was incompatible with the idea of the atom as a homogeneous sphere in which the electrons are embedded like sultanas in a cake, since such a structure would not be able to deflect positively charged particles from their paths. In the years from 1909 to 1911, Rutherford developed his own model of the atom, in which electrons orbit a very small nucleus of positively charged particles. By means of the distribution of the particles reflected on the gold foil and measured on the photographic plate, Rutherford was able to determine the size ratios in the atom. His observations showed that by far the largest part of the mass had to be located in a nucleus whose size, however, would occupy only a minuscule part of the entire atom.

Because this atomic model was also based entirely on the rules of classical physics, fundamental problems were not far away: What could keep the negatively charged electrons from crashing into the positively charged atomic nucleus? In the solar system, the centrifugal forces of the planets are in balance with gravity. But in the atom, according to Maxwell's theory, the negatively charged electrons would have to constantly emit radiant energy during their circular motion, gradually slowing down and crashing into the atomic nucleus within a fraction of a second. Apparently, nature did not adhere to Maxwell's equations.

Around 1910, classical physics had finally reached a dead end. For twenty or thirty years, physicists had made astonishing progress and yet had achieved nothing more than dismantling classical physics. There was no longer a consistent world view. The well-established order of physics lay in ruins.

3

Mathematics Becomes Paradoxical—Georg Cantor and the Insurmountable Contradictions of the Infinite

Just as physics slid into a fundamental crisis at the end of the nineteenth century, from which it would not recover until well into the twentieth century, mathematics suffered the same fate. The triggers for these shock waves had the same origin. While physicists were despairing of the world of the very small, because they were getting nowhere with the question of whether atoms exist and what their properties could be, mathematicians found they lost their footing as soon as they approached the infinitely small or the infinitely large. In both disciplines, the preoccupation with ever more extreme quantities was leading to unsolvable contradictions.

The fact that the usual explanation of the world collapses as soon as one deals with the concept of the infinite was already known to the ancient Greeks. The paradoxes they handed down are mostly about a process being repeated an infinite number of times, resulting in smaller and smaller units. Even today, these contradictions confuse our senses because logic and experience no longer fit together:

- Zeno's turtle paradox describes how a turtle gets a small head start in a race with the speedy Achilles. As soon as Achilles reaches the turtle's starting point, the turtle has already moved a little further. When Achilles reaches this point, the turtle has already moved a little further again, even if only by an even smaller distance. And so it goes, on and on. *Logically*, in an infinite series of "steps", the distance between Achilles and the turtle becomes

L. Jaeger, *The Stumbling Progress of 20th Century Science*, https://doi.org/10.1007/978-3-031-09618-1_3

smaller and smaller without Achilles ever being able to catch up. In *our experience,* however, he will quickly overtake it.

- In the arrow paradox, a flying arrow occupies a specific, precisely outlined location at every moment of its flight. For an infinitely short period of time, the arrow is at rest at this location. *According to logic,* the arrow must be at rest overall, since it is also at rest at each of the infinitely many and infinitely short instants of time. But *our experience* tells us that it is flying.

- The question of whether or not matter is divisible to infinity also leads to a paradox. If it is divisible, there must be tiny spaces in it along which it can be divided. In this case, if matter is divided an infinite number of times, it will end up consisting only of "empty" division lines, i.e., of nothing. This contradicts our experience that matter exists. So ultimately there must be indivisible particles. On the other hand, doesn't an atom, however small it may be, have to occupy a certain region of space? But then smaller spaces than the one it occupies are always conceivable. And since we can mentally divide space further and further, atoms can no longer be imagined as indivisible. Indivisible particles cannot therefore exist.

The Greek scholars could not resolve the inner contradictions of these thought games. Some of them, including Zeno and Parmenides, trusted logic rather than experience and concluded that motion and matter were only appearances. For over 2000 years, the infinite remained a concept that neither philosophers nor mathematicians knew much about.

The third paradox, which ultimately deals with the existence or non-existence of atoms, remained unsolvable into the twentieth century. After classical physics had failed because it could not explain those phenomena in the world on the smallest scales, there was no new foundation for many years that could have supported a better theoretical edifice. However, the first two paradoxes, the race with the turtle and the flying arrow, could be solved with the help of a completely new form of mathematical calculation that was invented from the end of the seventeenth century.

The Forbidden Door

Modern mathematics initially developed in close connection with physics. The infinitesimal calculus (integral and differential calculus), invented independently of each other by Newton and Leibniz around 1680, initially served to describe physical phenomena and other concrete issues, especially motion.

To say that a car travels 50 km/h is only an approximation, because sometimes it travels a little slower, sometimes a little faster. With the infinitesimal calculus, it was now possible to describe the motion within any number of arbitrarily small, so-called infinitesimal sections, and thus to state the relevant instantaneous speed exactly. Physicists were satisfied because integral and differential calculus proved to be accurate enough to solve many physical problems that had previously caused them headaches. But there was no explanation satisfactory to mathematicians in the form of a clear proof of why this worked. As long as a mathematically coherent description of infinitely large and infinitely small quantities was lacking, infinitesimal calculus was no more than a tool with no connection to the fundamental laws of mathematics. In attempting to integrate integral and differential calculus into the fundamental laws of mathematics, mathematicians passed through the doorway that led to these infinities, a path that scholars had avoided as much as possible since antiquity.

Potential infinity is exemplified by the natural numbers (1, 2, 3, ...). For every natural number, you can always find an even larger one, which is why there is no natural number that is itself infinite. The ancient Greek mathematician Euclid proved that the set of all prime numbers also has this property.

Actual infinities are characterised by the fact that they cannot be generated using a finite algorithm. This is because the algorithm only ever produces numbers that consist of a finite number of characters. An example of an actual infinity is the set of real numbers. There is no algorithm that produces the whole set of real numbers.

The sentence *Infinitum actu non datur,* the actual infinite does not exist, goes back to Aristotle. He had already recognised that what mathematicians call "actual infinities" entail paradoxes. For him, there was only the "potentially infinite", by which he meant a set of objects whose counting can be continued forever, but only *potentially* has no end. For everything that can be counted can basically only form a finite set.

So, for a long time it was agreed that:

- It must be possible to construct everything infinite from the finite; it must therefore be possible to count it on the basis of the natural numbers.
- There is nothing *absolutely* infinite in nature.

Later, religion adopted another interpretation of the infinite: God alone is truly infinite. So, for both groups, philosophers and theologians, there were good reasons to give actual infinities a wide berth. However, it was already suspected that there could be different kinds of infinity.

It was precisely the desire to give the infinitesimal calculus a solid mathematical foundation that led mathematicians to deal with the infinite without blinkers—in this case, with the infinitely small. To this end, in his work *Cours d'Analyse*, the Frenchman Augustin-Louis Cauchy suggested in 1821 that two numbers could be infinitely close to each other, without coinciding. He also showed that, for a sequence of numbers that fulfilled certain criteria, the infinitely many elements making up such a sequence could sum up to a *finite* value. He thus validated the concept of limit, which can be used to define the most important concepts of infinitesimal calculus:

- continuity specifying functions without jumps,
- differentiability defining functions without kinks,
- integrals used to calculate areas within curvilinear bounds.

With the concept of limiting values, the paradox of continuous motion (arrow paradox) and that of infinite sequences with ever decreasing increments (Achilles paradox) could finally be solved: both sequences can be broken down into infinitely small intervals, but they also have definite limiting values:

- The arrow travels an infinitely short distance in an infinitely short time, and yet it covers a finite distance in a finite amount of time.
- The infinite sum over the ever-decreasing distances from Achilles to the tortoise also converges to a finite limit: the point in time when Achilles actually overtakes the tortoise.

These applications were relatively tangible, but more and more mathematics was now detaching itself from physics, which until then had remained rather easy to understand intuitively and visualise, and was also moving away from practical applicability. The further mathematicians ventured into abstraction, the more mysterious everything became and the more the assumptions of classical mathematics that were previously believed to be secure were called into question. For example, on closer inspection, mathematicians found that the methods for calculating lengths of curves, areas of surfaces, and volumes of regions of space were imprecisely defined from a mathematical standpoint. Everything turned out to be much more complex than the known, simple formulae suggested.

More Major Challenges

In 1854, the German mathematician Bernhard Riemann gave the calculus of integration the form we know today by introducing upper and lower bounds to the area under the graph of a function for ever shorter, but ever more numerous intervals. For example:

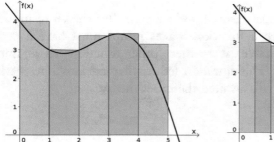 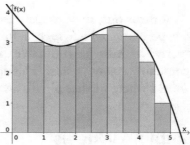

Upper sum with relatively large interval   *Subtotal with shorter interval*

The shorter the intervals become, the more closely the upper and lower sums approach one another—and the better they describe the actual value. Cauchy had already established similar sums, but Riemann now asked himself under what precise circumstances these sums would converge to clearly defined values. And, of course, when they would not. Mathematicians discovered functions with properties that defy intuition. For example, there are functions that are continuous everywhere but not differentiable at any single point—no one had expected that. Riemann himself began to work on functions with "jumps".

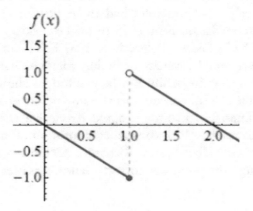

- He proved that every continuous, i.e., "non-jumping" function is integrable, i.e., one can calculate the area lying under the curve of its graph.
- Riemann was able to show that functions that are discontinuous, i.e., "jump", at only *finitely many* points, can also be integrated.
- The question Riemann had to leave open was: Are there also functions that are not continuous at an *infinite number* of points and yet are integrable?

As it turns out, such special cases do indeed exist. But the set of infinitely many points at which such a function makes jumps must have a certain property: it must be countable. Were there other functions that jump at uncountably many points? This question led mathematicians into a whole new field where further pitfalls awaited them: set theory.

Rumbling Wheels

The elementary concept of quantity goes back to the ideas of Aristotle. He mainly referred to finite quantities, which are easy to understand. In passing, the ancient philosopher also dealt with *potentially infinite* sets. On the other hand, a set that actually contains an infinite number of objects in every observation, i.e., contains a number of elements that cannot be counted with the natural numbers, should be impossible according to Aristotle. It was only in the course of the infinitesimal calculus that one began to deal with these *actual infinities*. Even those who had brought infinitesimal calculus to fruition, namely, Newton, Leibniz, Bernoulli, and Euler, found these infinities uncanny.

The very statement that different forms of infinity could exist threw mathematicians into great confusion. The German mathematician Georg Cantor realised that entirely new principles had to be developed to describe *how* infinite an infinite set is. Born in 1845 in St. Petersburg, Russia, he began his studies at the Swiss Federal Polytechnic, now ETH Zurich. He was soon drawn to the prestigious University of Berlin, where he habilitated. How he would have liked to stay in Berlin! But he was too far ahead of his time. He published a proof that a square contains as many points in its area as on each of its bounding lines. His mentor Leopold Kronecker tried to prevent the publication of this result, which contradicted intuition. Only after the intervention of a mathematician friend, Richard Dedekind, was Cantor's proof published. Although the proof was correct, Cantor was met with a barrage of angry reactions. He now had no chance in the capital, and for the rest of his life he had to be content with the provincial university in Halle.

Cantor remained faithful to infinities. He wanted to know, for example, whether there are more rational numbers than natural numbers. Even if both sets are infinitely large, the answer seems clear at first: there must be more rational numbers than natural numbers, because between two consecutive natural numbers there are many rational numbers; it can even be easily proved that between any two natural numbers there are an *infinite number* of rational numbers. The infinity of the rational numbers seems therefore to be infinitely greater than the infinity of the natural numbers, because Euclid had already said in his 5th axiom that the whole is always greater than its parts. But this is exactly where we fall into a trap. A simple mathematical proof shows that there are actually *no* more rational numbers than natural numbers.

Natural numbers ℕ: positive integers, i.e., 1, 2, 3, 4, 5, ... Depending on the definition, zero may also be included in this set.

Rational numbers ℚ: all numbers that can be represented as a ratio of two whole numbers, for example, $1/2$, $2/3$, $5/5$, etc. All natural numbers are also rational numbers.

The simple trick Cantor used to show that every natural number can be assigned to exactly one rational number and vice versa—that there are no more rational numbers than natural numbers—had, incidentally, already been used by Galileo to prove that there are just as many square numbers as natural numbers. He came to this insight by means of a paradox that goes back even further in history—to Aristotle. He had carried out the following thought experiment.

Two wheels are attached to the same axle. One has twice the radius of the other and therefore twice the circumference. The small wheel is glued flat to the big one, so when the big one turns, the small one turns with it without slippage. Because of the different sizes, only the big wheel touches the ground.

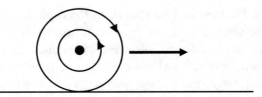

Now this combined wheel rolls exactly one full revolution of the big wheel. The small wheel moves with it because it is attached to the big wheel. What distance does any point on the smaller wheel travel? The small wheel, like the big wheel, has made one full revolution. Since it only has half the circumference, it should also only have travelled half the distance, and yet at the end the small wheel is in the same place as the large wheel. How does that fit together?

Galileo Galilei proposed an elegant mathematical solution to this paradox. For this, he first imagined the two wheels as hexagons. When rolling the large hexagon, one obtains a continuous line whose length corresponds to the circumference of the hexagon. The line that the small hexagon leaves behind, on the other hand, is interrupted by several gaps, because it has to "hop" at the corners to keep up with the large hexagon. Galileo now mentally increases the number of corners more and more, first to many dozens, then to 100,000, up to infinity. Thus, the gaps on the path of the small wheel become infinitely small, but also infinitely numerous, so that the distance actually rolled becomes as long as the circumference of the large wheel.

Galileo continued as follows. The path traced by the inner wheel consists of infinitely many points separated by infinitely many, infinitely small empty spaces, while the path traced by the outer wheel consists entirely of the points with no empty spaces between them. Should not then the path described by the outer circle contain a greater infinity of points than that of the inner circle? No, because every point and every distance on the inner circle can be assigned exactly one point and one distance on the outer circle. To do this, we only need to draw the radii from the center to the outer circle. Each passes through exactly one corresponding point of the inner circle. So, we have two infinities that are different according to our intuition, but mathematically the same. The geometric consideration with the two wheels can be transferred to many questions that concern countable quantities:

- Galileo used it to show that the number of natural numbers and the number of square numbers are equal.
- Cantor proved that the number of natural numbers and the number of rational numbers are equal. This only works because the rational numbers

are "countable", i.e., they can be put in a one-to-one correspondence with the natural numbers.

- Similarly, when we ask whether there are more natural numbers (1, 2, 3, 4 ...) than even numbers (2, 4, 6, 8 ...). Intuition tells us that there must be half as many even numbers as natural numbers. But even in this case, mathematical proof says otherwise: the set of even numbers is *not* smaller than the full set of natural numbers.

With the real numbers, however, something quite different became apparent. Cantor proved that they are *not* countable and that their number is actually infinite—that is, actually infinite beyond any human imagination. Cantor had thus shown for the first time that, *mathematically speaking, there are* also infinite quantities of different sizes; the real numbers are "even more infinite" than the rational numbers. No wonder the ancient Greek mathematicians had such a hard time with the existence of irrational numbers like $\sqrt{2}$. Mathematicians today say that the (for human understanding infinitely large) measure of all rational numbers within the real numbers is zero. For even if there are an infinite number of rational numbers, their number within the set of real numbers is infinitely small.

Irrational numbers are numbers that cannot be represented as fractions. Examples are π and $\sqrt{2}$. Irrational numbers have an endless decimal expansion. Therefore, there is no (finite) algorithm that can produce every real number. This is because algorithms can only produce numbers that can be represented with a finite number of digits; this is the Achilles' heel of many encryption techniques, for example.

Real numbers $\mathbb{R}$: All rational and irrational numbers together.

The insight that the number of real numbers is actually infinite was a breakthrough. For the infinitely small, which is the basis for the infinitesimal calculus, also possesses this kind of infinity. Without the actual infinity of the real numbers, there would be no infinitesimal calculus, and without infinitesimal calculus, there would be no mathematical physics! Only with the introduction of the actually infinite could sequences of limiting values, as introduced by Cauchy and Riemann for differentials and integrals, be defined perfectly and without contradiction. Formulated mathematically, it is only in the set of real numbers that every Cauchy sequence (a sequence of numbers in which the distance between the sequence members becomes arbitrarily small in the course of the sequence) has a definite limiting value. This would not work in the sets of natural or rational numbers. Thanks to these actual infinities, the infinitesimal calculus now stood on a completely firm mathematical foundation.

The Lies of the Cretans and the Truth About Infinities

Cantor had introduced an important property of infinite sets: their size, also called their cardinality. If the elements of two sets can be mapped unambiguously onto each other, then the sets are "equal in size". The natural numbers and the rational numbers have the same power, as do the *even natural* numbers and *all* natural numbers. But the power of the set of real numbers is even greater. Cantor showed precisely that the real numbers are *not* countable, i.e., cannot be mapped one-to-one onto the natural numbers, and thus introduced the actual infinite into mathematics. He also realised that there even exists an infinite hierarchy of increasingly larger sets—an infinite sequence of infinities! For this, one only has to consider the sets of all subsets of the real numbers, and then the set of all subsets thereof, which are called power sets by mathematicians, and so on ("infinities", "infinite infinities", "infinitely infinite infinities," etc.). Cantor was able to show that the power of any set is smaller than the power of its power set. For finite sets this is self-evident, but for infinite sets it still had to be proved. Today, this theorem is called Cantor's theorem.

> The **cardinality of a set** is defined as the number of elements in the set.

Faced with so many infinities, Cantor's head was spinning. Against all odds, he persevered with his findings, believing that he was on the way to finding out about God himself. Between 1879 and 1884, he published a series of six articles that formed an introduction to his general set theory. But the opposing camp, led by his former teacher Leopold Kronecker, continued to vehemently resist actual infinities. Mathematical concepts should continue to be admissible only if they could be constructed from the natural numbers in a finite number of steps.

Mathematics was now in a profound crisis, triggered by these questions about infinities. Cantor had not committed any crime, after all—even if the established professors were still reluctant, actual infinities were out there in the world. Here, an almost contemporaneous parallel to the events in physics becomes clear: the landscape of traditional mathematics with its potential infinity no longer satisfied the demands, but no one had yet found a firm foothold on the new territory where actual infinities have their home.

The bitter resistance of the experts to the hierarchy of infinitely many infinities introduced by Cantor had a deeper philosophical reason: although it could ultimately place the infinitesimal calculus on a solid foundation and thus finally resolve certain paradoxes, it opened the door to further paradoxes that called into question the validity of mathematics as a whole. In fact, Cantor did not succeed in placing his hierarchy of the infinite on a mathematically sustainable foundation, because he landed exactly where the ancient philosophers had already been shipwrecked: actual infinities seem to be simultaneously different and equal! This logical problem corresponds to the liar's paradox, already known to the ancient Greeks. Here is the most widely known version:

Epimenides the Cretan says: "All Cretans lie."

Epimenides' claim can be true or untrue, and in each case, he is a liar and at the same time not a liar. No matter how you turn it, the sentence always ends in an irresolvable contradiction.

Mathematicians went round in circles in a similar way when they dealt with the cardinality of the largest infinite set. It was clear that there is no largest cardinality of sets, i.e., no largest set, because the power set of that set must have an even larger cardinality. If we now add Cantor's theorem ...

The power of any set is less than the power of its power set.

... an insoluble contradiction arises. For if one considers the set of *all sets,* it would also have to contain its power set. This power set would simultaneously lie inside and outside the set with the greatest cardinality.

The British mathematician Bertrand Russell put this contradiction into another form in 1901. He said:

Consider the set of all sets that do not contain themselves.

If this set is not itself an element of itself, then this sentence says that the set must contain itself. But if it contains itself, then it falls outside the definition as the set of all sets that do *not* contain themselves.

Since set theory is the basis for all other branches of mathematics, the paradox of Cantor and Russell threatened the entire subject. For in logic, it is true that every statement can be proven from a contradiction. The connection can be clarified—not quite cleanly from a technical point of view, because logical and linguistic elements are mixed—in this way: Someone says: "If 1 + 1 = 3, then elephants can fly." If the premise is already wrong, the second half of the sentence can contain anything. So, if something is wrong in set theory, then there are only flying elephants in mathematics. It was a disaster! After all the arguments about the different infinities, the Liar's Paradox in the form of Cantor's and Russell's theorems plunged mathematics into a fundamental crisis.

New Construction Instead of Renovation

To put an end to the deep uncertainty, the German mathematician David Hilbert attempted to redefine the whole of mathematics from scratch. Starting from an agreed set of axioms, he wanted to establish the foundations of mathematics step by step with logical proofs. The number of these steps was to be finite, or in a weakened variant, countably infinite. On 8 August 1900, at the International Congress of Mathematicians in Paris, he presented the ten most important unsolved problems in mathematics. These were the remaining stumbling blocks on the way to a coherent mathematics. Shortly afterwards, he extended this list to a total of 23 problems from the sub-areas of geometry, algebra, number theory, logic, topology, arithmetic, analysis, etc. This list went down in the history of mathematics as the "Hilbert Programme". It led to the elaboration of modern proof theory in mathematics in the 1920s and remained seminal for the development of mathematics in the following decades. Most of Hilbert's problems have been solved to this day, with only a few remaining open. The most famous among mathematicians is the eighth problem: the Riemann conjecture for the zeros of the zeta function, but that is a completely different construction site.

Axiom: a statement that does not require proof. Starting from the axioms, mathematical theories are derived by deduction. The prerequisite is that all axioms should be free of contradictions – this was precisely the problem with set theory.

In second place on Hilbert's list of 23 challenges was the formulation of a non-contradictory set theory, i.e., the resolution of the Liar Paradox that results from dealing with actually infinite sets. Only then would all other theories of mathematics be officially valid. But to do this, the matter had to be revisited from scratch, starting with the question, which extends into philosophy, of how one should define the existence of mathematical objects in the first place. What exactly *is* a set? Two camps formed here:

- On the one hand, David Hilbert with his view that mathematics is universal—i.e., valid even without the existence of human thought—and that its propositions can therefore be proven or falsified without contradiction. If something is not right, it is wrong, and if something is wrong, it is not right. There is no third option.
- On the other side, the Dutch mathematician Bertus Brouwer with the position of so-called intuitionism: mathematics is exclusively the result of the mental activity of people, and there are no objectively real principles in it. In this system, only what has also been directly and unambiguously proven is considered true, and this by means of explicit construction. Thus, the law of excluded middle loses its rank as a method of proof—the truth of a statement can no longer be proven by confirming it as "not false".

Before the concept of intuitionism existed, the most important mathematicians—Henri Poincaré, Henri Lebesgue, and Émile Borel in France and the highly influential Nikolai Luzin in Russia—were already followers of this school of thought. Intuitionism also fits with a remark by Kronecker, Cantor's opponent: "God created the natural numbers; everything else is the work of man." The advantage of this approach to mathematics is that many statements from set theory are omitted, since they were only proven by showing that they were "not false", and therefore no disturbing paradoxes occur.

The law of excluded middle: In classical logic, A is either true or false. There is no third option. In mathematical notation:

$$A \vee \neg A$$

(A or non-A)

In Hilbert's classical mathematics, on the other hand, paradoxes arise because here the validity of mathematical proofs is less strictly regulated. Within its framework, for example, it is possible to work with abstract sets without specifying any means of finding and constructing them. The hurdles that must be overcome to call a mathematical system complete or consistent are thus relatively low:

- If either a "true" or "false" can be derived for all statements within the system, the axiom system is called *complete*.
- It is also enough for the system to have no internal contradictions, i.e., no statement exists that can be proven to be "true" and "false" at the same time. As long as such a proof is missing, the system is said to be *consistent*.

Mathematical system: Part of mathematics, for example algebra or set theory.

Hilbert's mathematics was, so to speak, the freestyle, while Brouwer's was the duty. In mathematical practice, the two ran parallel to each other: the theories à la Hilbert made use of the more sophisticated intuitionistic concepts à la Brouwer, while on the other hand, using methods à la Hilbert, a number of important results on the logic of intuitionism were found. Meanwhile, the question of whether one may equate "true" with "not-false" and "false" with "not-true" remained unresolved.

The Big Bang

In 1908, the German mathematician Ernst Zermelo proposed a new version of set theory that tried to avoid the paradoxes of Cantor's existing set theory within Hilbert's system. The latter was now called "naive set theory", since in the eyes of Zermelo and his colleagues it was more descriptive and not strictly logical. Zermelo replaced Cantor's understanding of sets with modifications of this theory proposed in the 1920s by Abraham Fraenkel and Thoralf Skolem in addition to Zermelo. They led to an axiomatic set theory which today is abbreviated as "ZFC", which stands for Zermelo–Fraenkel–Choice. Here, "Choice" refers to a certain axiom introduced by Zermelo, known as the axiom of choice. Once the latter was recognised as indisputable, this theory became widely accepted as the standard. The axiom system ZFC is still the basis of set theory and thus of mathematics today.

However, there remained the first and most important problem on Hilbert's list, which Cantor pursued with great commitment until the end of his life: Are there other infinities whose power lies between the uncountably infinite set of real numbers, which is so important for the description of nature by physics, and the countably infinite set of rational numbers? He tried to prove that there is *no* such set. Finding a proof for this "continuum hypothesis" was also the most burning problem for Hilbert. For the consistency of the Zermelo–Fraenkel set theory stands and falls with it, and so does the mathematicians' claim to describe reality.

And again, a strange parallelism between the events in physics and mathematics becomes apparent. In both sciences, complete, complex, and coherent explanations of their internal relationships had been found, and in both, individual pieces of the puzzle end up calling everything into question. In physics, it was the discoveries that the speed of light is constant and that light can be a particle and also a wave. In mathematics, the situation was even more dramatic: it turned out that, on the basis of ZFC set theory, the continuum hypothesis can be neither refuted nor proven definitively! In addition to the categories "True" and "False", there was suddenly a third category: "You will never know"!

How did this landslide come about? Together with his colleague Wilhelm Ackermann, Hilbert published the work "Grundzüge der theoretischen Logik" in 1928. In it, they posed the question: Are the axioms of a formal system—in this case, ZFC set theory—sufficient to derive every statement that is true within the system in a finite number of proof steps? Hilbert answered this question categorically in the affirmative. As late as 8 September 1930, in his retirement speech to the Society of German Scientists and Physicians in Königsberg, he reiterated his position on the possibilities of mathematical knowledge.

Exactly one day earlier, at the same conference, Kurt Gödel, a 25-year-old student, had announced the forthcoming publication of his "incompleteness theorem" at a round table with friends and colleagues. Gödel's announcement attracted little attention. Only John von Neumann, who will be discussed in a later chapter, pulled him aside to talk. Hilbert and Gödel, on the other hand, did not meet. It was only after Gödel published his thoughts in 1931 under the title "On Formally Undecidable Theorems of Principia Mathematica and Related Systems" that mathematicians recognised the significance of his work.

Reminder: A mathematical system is **complete** when it proves either the correctness or the negation of each proposition.

> It is **consistent** when there are no contradictions in it.

In fact, Gödel had already proved in his doctoral thesis of 1929 that in sufficiently complex axiom systems (such as the arithmetic of the real numbers and also Cantor's and ZFC set theory), consistency and completeness are generally mutually exclusive. In such systems, therefore, there are always propositions that are neither provable nor disprovable with the means of this system; the systems are *in principle* incomplete. Gödel's first theorem reads accordingly:

> If an axiomatic formal system is consistent, it cannot be complete.

This means that, in contradiction-free systems there are always also unprovable statements. The proof of this first Gödel theorem goes back to the liar paradox of Epimenides. Now the Cretan Epimenides no longer says "All Cretans lie", but:

> The proposition G reads: "This proposition is not provable in the system F."

The contradiction is the same: if G is true, it is simultaneously false; if G is false, it is simultaneously true.

With Gödel's theorems, attempts to find a satisfactory basis for mathematics ended after half a century. For Gödel's theorem applied both to the continuum hypothesis and to the axiom of choice within the ZF system. Mathematicians had to come to terms with the fact that ...

- ... mathematics must always remain open because of its own structure and logic,
- ... some of the Hilbert problems are fundamentally unsolvable,
- ... the framework of Cantor's as well as higher-order set theories does not allow an answer for some questions, in the sense that the corresponding hypotheses can neither be proved nor disproved,
- ... mathematics now contains a third category alongside "true" and "false", namely, "undecidable in principle",
- ... structurally, the human mind has to live and make do with uncertainty and ignorance.

What a shattering of faith in the purity and beauty of mathematics! The idea of eternal progress towards ever higher knowledge was now gone once and for all, even in mathematics. Hilbert's dream of a complete and consistent mathematics had been shattered.

However, Gödel's incompleteness theorems do not completely rule out the possibility that the consistency of a theory can still be proven by completely different means. For Gödel's second theorem reads:

> The consistency of axioms within an axiom system cannot be proven within their own system.

Thus, undecidability only means that the truth or falsity of a statement cannot always be proven *within* the system under consideration. If a larger theoretical framework is used, everything is open again. Whether there are "once and for all undecidable" statements whose truth value can never be known was not the subject of Gödel's considerations. Thus, Gödel was already able to show that it is impossible to prove within ZFC that the continuum hypothesis (CH) does not hold. In 1964, the American mathematician Paul Cohen then showed that it is equally impossible to prove the validity of the CH from this. Now, one could add new axioms to the ZFC axiom system that would clearly show whether the continuum hypothesis is true or false. But which axioms are suitable for this? Mathematicians are still trying to answer this question today.

Where do we go from here? Mathematicians have been asking themselves this question ever since they discovered that the criterion "undecidable" exists and that for some problems no definitive "true" or "false" can be found even after a very long search. Do we just have to research a little more persistently or is there really no solution?

Is It Worth the Effort?

Mathematicians went so far into abstraction that they shook the foundations of their subject. Many of them also paid a high price personally. The fact that people who revolutionise thinking mentally and physically go to their limits, and not infrequently beyond them, had already been illustrated by the example of the physicist Boltzmann. Like him, Cantor suffered from phases of mental instability, exacerbated by the oppressive resistance of his professional colleagues, which never abated. Severe depression was an increasing him, and he repeatedly had to seek longer and longer periods of psychiatric treatment. In January 1918, he died in the closed ward of a clinic.

Gödel's fate was also tragic. He was highly paranoid and depressive. When his close friend and teacher Moritz Schlick was murdered by an Austrian Nazi in 1936, Gödel suffered a nervous breakdown and thereafter had an obsessive fear of being poisoned. He only ate food that his wife Adele had prepared for him. At the end of 1977, Adele was hospitalised for a long period, and refusing to eat, Gödel starved to death.

Was this excursion into abstraction just an aberration? *Mind-blowing* in the truest sense of the word and yet basically completely useless? Surprisingly, many concepts of abstract mathematics later turned out to be very fruitful for theoretical and thus also for applied physics. Gödel's abstract mathematics, which was understood by only a few experts worldwide, even made it into the concrete world of the new machines that were soon to conquer the globe: computers.

It was foreseeable that computers as logical machines would eventually encounter the same problems as mathematical logic. Six years after the publication of Gödel's incompleteness theorems, the English mathematician Alan Turing—whose achievements will be comprehensively acknowledged in a later chapter—"translated" its abstract logical language into a comparatively simple algorithmic mechanism. He was concerned with the so-called decision problem, which Hilbert had already described—in a roundabout way—as the tenth problem. What Turing was looking for was a method for deciding, on the basis of logical rules, whether a certain statement was provable from the available axioms (Hilbert specifically asked for a method that decides whether any given diophantine equation is solvable or not). The idea was that, using such a method, one would never have to work on a problem for many years when there might actually be no prospect of success—for example, the problems relating to ZFC discussed above.

In fact, the decision problem is a special form of the problem of completeness and consistency of axiomatic systems. Turing transformed it into a new form known as the halting problem: if a machine works on a complex problem for a very long time, we cannot know whether:

- the problem is very difficult to solve and therefore requires a lot of time, or
- it is actually unsolvable and therefore the machine will never come to an end.

Turing came to the realisation that there is no general logical algorithm that can reliably tell us whether a computer, when working on any problem, will at some point reach a definite end and thus an answer. The calculating machine he designed for this purpose works as follows:

- Symbols from a finite alphabet are mapped on an infinitely long strip of tape divided into individual cells. At any given time, a read and write head is always located on exactly one of these cells and reads the character on it.
- The read and write head also has its own internal state in the form of a finite register. After reading, it changes the character in the cell according to a predefined rule table. Depending on the character read and its own internal state, it writes a new symbol into the cell according to a finite table with uniquely specified instructions, the decision matrix, and at the same time moves itself into a new state of its own.
- Then the machine moves either one cell to the left or right on the tape and repeats the process with the new cell and the new state of the read and write head. If there is no entry in the decision matrix for the combination of the cell symbol and the internal state of the read and write head, the machine stops.

The mathematical machine constructed mentally by Turing to execute this mechanism is known today as a "Turing machine". With its help, Turing succeeded in 1936 in showing that any general algorithm that is supposed to decide whether, for any programme (decision matrix) and any tape input, the programme stops at some point or whether it continues to run forever must contain a self-contradiction. So there can be no such algorithm. He thereby made two important discoveries:

- Just like Gödel before him, Turing had proved that Hilbert's general decision problem could not be solved.
- Even before the first computer was built, Turing had defined the functioning of all modern computers with his machine and at the same time discovered their theoretical limits.

But here too it is only *within* the system that we are forced to say: "It cannot be decided; we will never know." Within the rules of the computer world we know today, in which digital states are processed one after the other with a finite number of characters, we will never know whether a calculation will come to an end. In a higher-level system, however, the question would be open again. What such a system might look like is currently inconceivable. Even quantum computers, which no longer work only with the states 0 and 1, but in principle with an infinite number of states at the same time, will not bring about a system change. They may work billions of times faster than classical computers, but they will not be able to solve problems such as the decision or stopping problem. There will probably only ever be one answer to them: we will never know.

4

Darwin's Hesitation and Mendel's Diligence—Life as a Plaything of Atomic Elements

Until the eighteenth century, it was taken for granted by those in Christian cultural circles that the Earth, together with everything that lives on it, had come into being in a single act of creation. According to this creationist view, the world had remained the same ever since and would not change until the Last Day. When the Irish theologian James Ussher calculated in 1650, on the basis of the time periods given in the Bible, that the Earth had come into being on 23 October 4004 BC at 9 o'clock in the morning, this date was well received by his contemporaries and considered plausible.

Geologists were the first to suspect that this date and the statement in Genesis that the Earth had been created in seven days did not quite fit with their observations. Was it really possible that the Earth with its complex rock formations was as young as Ussher had calculated? And how did it fit into the picture of a never-changing world that fossilised shells were found in some mountainous areas, far from any sea? Many other fossil finds, which were obviously the remains of animals and plants that could no longer be found on Earth, caused great confusion.

A first alternative origin story for the Earth was developed by the Frenchman Georges-Louis Leclerc, Comte de Buffon, in the last third of the eighteenth century. He imagined that the existence of the Earth was due to a cometary impact on the Sun. According to him, our planet formed from some of the solar matter ejected into space during this event. In order not to stray too far from the processes described in the Bible, he defined seven stages in which the ball of fire cooled to a hard sphere, atmospheric gases condensed,

L. Jaeger, *The Stumbling Progress of 20th Century Science*, https://doi.org/10.1007/978-3-031-09618-1_4

and huge oceans formed, in which—as the frequently found fossils of marine life suggested—life arose. In a further step, the land masses were formed by violent volcanic eruptions. Finally, as the crowning achievement of Earth's history, man entered the stage. According to Leclerc's calculations, this story would have required at least 100,000 years.

Based on this view, two opposing geological theories developed:

- The German geologist Abraham Gottlob Werner believed that all rocks were formed by sedimentation in an ocean created by a flood.
- The Scotsman James Hutton, on the other hand, agreed with Leclerc, advocating the thesis that the essential formative force was a central fire inside the Earth and that the continents owe their existence to volcanic forces.

Today we know that both sedimentation and volcanic activity have shaped the continents. Hutton also recognised a third important force shaping the land masses: erosion. A fourth force, plate tectonics, was discovered much later.

Hutton assumed that the shaping of the Earth must have taken place even more slowly than Leclerc had thought and—this was a completely new idea—that everything continued to be in a constant flux. Geologists began to use scientific methods to search for possible mechanisms underlying these processes. At the beginning of the nineteenth century, the Scottish geologist Charles Lyell further developed the thoughts of his compatriot Hutton. A copy of his book "The Principles of Geology: An Attempt to Explain the Changes of the Earth's Surface by Forces Still at Work" went on a world tour in 1831 on the HMS Beagle. The book's owner was Charles Darwin.

Turtles and Mocking birds

Charles Darwin was supposed to follow in his father's footsteps and become a doctor, but he was much more interested in biology, and initially in botany. Like many natural scientists of his time, he went to excessive lengths to collect, identify, and preserve all possible representatives of the plant world in order to advance the inventory and systematisation of the immense wealth of species in the world. When he dropped out of medical school, his father was disappointed and decided that Charles should study theology and become an ordained priest. But Darwin junior also pursued this study with only moderate zeal. He preferred to devote himself to his newly discovered love of insects and attended the lectures of the botanist John Henslow, with whom he soon formed a deep friendship. Henslow encouraged Darwin to join a geological expedition to Wales in 1831. This trip was a key experience for Darwin. In August of the same year, Henslow arranged for him to have the opportunity of a lifetime: Captain Robert FitzRoy was looking for a young man with the best education as a scientific companion on a surveying voyage along the coast of South America that would last up to five years. The budding theologian Charles Darwin succeeded in gaining his father's approval after initial rejection, and on 27 December 1831 the Beagle set sail with the 22-year-old naturalist on board.

Reading Lyell's book had made Darwin an ardent supporter of the latter's idea that geological conditions on Earth evolve steadily and over long periods of time. On his journey, he found ample evidence to support this theory. Witnessing a volcanic eruption and a strong earthquake, he was even able to observe directly how the Earth's surface could be deformed. Darwin, however, did not go down in history as an important geologist. His merit was the discovery that there are developments, not only in the world of rocks, but also in biology.

The Beagle docked at numerous places on the South American continent and each time Darwin penetrated far into the interior on horseback or on foot, crossing deserts, climbing mountains, exploring forest areas, wading through rivers, and collecting everything he could lay his hands on: rocks, plants, insects, reptiles, fish, birds, and mammals. He identified and catalogued them, observed the animals' behaviour, studied their geographical distribution, and compared their characteristics with those of similar species from neighbouring regions. The most significant stopover for Darwin's later theory of evolution was made by the Beagle on the Galapagos Islands in autumn 1835. Here he found laboratory conditions, so to speak, under which the animal and plant species that had made their way here had had to adapt to their respective environments:

- There had never been a connection between the islands formed by volcanic activity and the mainland. The ancestors of all living creatures on the Galapagos Islands had been carried there at some point by storms or as stowaways on flotsam.
- The nearest land mass, the coast of Ecuador, is a thousand kilometres away. So it was only individual creatures or small groups at a time that reached the Galapagos Islands and were given the opportunity to gain a foothold here.

- Although the islands are located close to the equator, they have a temperate climate due to an Antarctic ocean current. Living conditions are therefore very different from those on the tropical South American mainland, where the animal and plant species migrated from.
- The various islands themselves also provide different climatic conditions. For example, some islands are very dry, others rainy. Thus, each island provides a unique habitat for the plants and animals that live on it.

Two animal species in particular aroused Darwin's interest. One was the giant tortoises, typical of the islands (the Spanish word for a particular form of tortoise shell is "galápago"). The islanders drew Darwin's attention to the fact that the turtle shells differ slightly but clearly from one another depending on the island. On the other hand, Darwin discovered four mockingbird species on the Galapagos Islands which were very similar to the already known species on the North and South American mainland, but also differed clearly in some features. For example, they had learned to use the food available on the islands—including carrion and newly hatched turtles—as a food source. Their breeding behaviour was also quite different from that of their relatives on the mainland.

Darwin had not yet recognised the connection between the shape of the beak and the food source of the finches that would later be named after him, of which he identified 13 new species on the Galapagos during his journey—the British ornithologist John Gould discovered this later on the basis of the specimens Darwin had collected. Darwin had even neglected to record which of the islands the finches came from in each case; he did not make this mistake with the mocking birds.

In October 1836, the Beagle returned to England. Among its cargo was a wealth of taxidermy, skins, bones, and other finds collected by Darwin, as well as more than 2500 pages of notes that he now had to evaluate. He suspected that the specific characteristics of mockingbirds and tortoises were adaptations to the special conditions of their environment that had differentiated over time. This was a clear contradiction to the creationists' view that animals and plants should have immutable characteristics, precisely those with which they were created. Darwin's assumption was far from being scientific evidence. It was also completely unknown how such adaptations should take place.

Chance and Necessity

Two years after Darwin's return, the "Essay on the Principle of Population" by the English economist Thomas Robert Malthus gave him a decisive and thought-provoking idea. "Population has a permanent tendency to increase beyond the measure of the food available", Malthus had written in 1803. Since the exponential growth of the human population inevitably exceeds the amount of available resources, a limitation of population growth was pre-programmed by hunger, disease, poverty, and crime, all problems that were still omnipresent in eighteenth and nineteenth century Europe. Malthus concluded that, in human society, the fittest survive, while the weaker parts of the population perish.

Malthus' views were already highly controversial at the time—and rightly so—but the British philosopher and sociologist Herbert Spencer went one step further. He called what Malthus described a "struggle for existence" and expanded it by saying that only those survive who can best adapt to the circumstances. Spencer called this *survival of the fittest*. Darwin developed the decisive idea for his theory of a natural selection process in the plant and animal kingdoms from these thoughts, which were aimed at human society: the individuals of a species compete for the goods that are necessary for survival but at the same time scarce, such as food, shelter, and protection from enemies. Individuals or species that can adapt particularly well to the constantly changing living conditions have a better chance of surviving. Species whose individuals cannot adapt to a longer period of drought, for example, become extinct. When Malthus and Spencer, used the expression "struggle for existence", they actually had in mind a situation in which the physically strongest and most ruthless would be victorious and survive, while Darwin interpreted the term "fit" as the quality of coping particularly well with the existing conditions, i.e., fitting in the other sense. Nevertheless, his name became the eponym for a brutal "Darwinian" struggle for survival.

But how does this adaptation work? Darwin recognised the obvious:

- The individuals of a species differ from each other, sometimes even significantly. There is a high variability within one and the same generation.
- There are also differences in expression between generations—offspring do not exactly resemble their parents.

The two effects together enable the variation of species-specific characteristics in the long term. Just as many small changes over a long period of time have shaped the appearance of the continents in the history of the Earth, the smallest changes from generation to generation have led to the current characteristics of the species. Darwin summarised the two basic principles of his theory of evolution in two words: variation and selection. They can be used to give the short version of Darwin's theory of evolution:

- **Variation**: Random change in the characteristics of individuals and thus their chances of survival and reproduction.
- **Selection**: Well-adapted individuals produce more offspring. Depending on whether a new trait is useful or harmful for the survival and reproduction of the individual, it spreads within the species or dies with the individual.

Darwin's theory offered a brilliant answer to a philosophical question which had caused disagreement since antiquity: are changes in the world caused by chance or by necessity? (It was Christianity that first brought the idea of an unchanging world into play.) According to Darwin, both are true. In variation, chance prevails; in selection, necessary principles come into play. His theory does without a conscious selection process and without a concrete goal, so the ordering hand of God becomes dispensable.

Darwin's Struggle for Existence

Darwin set out the essential ideas for a theory of evolution at an early stage. So why did he nevertheless need more than twenty years to publish his ideas in the form of a convincing theory? In fact, he was aware of the explosive nature of his ideas and feared the fierce discussions to be expected. Above all, he shied away from confrontation with strict believers in creationism; he already found it difficult to come to terms with the theory in the context of his own faith—and that of his deeply religious wife. So he withdrew in order to assemble all the scientific evidence into a finished and coherent framework. For many years he examined the individual characteristics of around 10,000 specimens from the animal and plant kingdoms. But even in possession of this immense store of data, which underpinned his concept of the origin of species, Darwin still hesitated to publish his findings.

Then, in the autumn of 1858, twenty-two years after the return of the Beagle, he received a letter from the naturalist Alfred Russel Wallace, describing a theory of the evolution of species that was similar in its basic features. Darwin was shocked and feared for the authorship of his ideas. He convinced Wallace to publish the results of their work together. Wallace included an outline of Darwin's theory in his article, which appeared in 1858. Darwin published his entire theory of evolution in a single work in November 1859: "On the Origin of Species by Means of Natural Selection, or the Preservation of Favoured Races in the Struggle for Life". The edition ran to 1250 copies and was completely sold out after only one day. Darwin's five main theses were at last made public:

1. **Theory of species variability**: Species are constantly evolving over time.
2. **Theory of descent**: Different species can be traced back to common ancestors.
3. **Theory of differentiation**: Species branch out over time, new species emerge, others die out. This differentiation is irreversible, in the sense that former species do not return.
4. **Theory of gradual variation**: The further development of species does not occur in leaps, but in small steps, triggered by random variations in single individuals.
5. **Theory of natural selection**: The evolution of species is driven by the competition of numerous individuals for limited resources. The variations lead to differences in survival probability and reproductive success.

Darwin's **first thesis**, for which he provided countless examples, was no surprise. Even before him, there had been evidence of species evolution, including numerous fossil finds. Animal and plant breeding had also long since shown that species can change considerably over generations. Darwin also supported his **second and third theses** with a large amount of evidence from fields as diverse as comparative anatomy, embryology, and geology. In sum, they were so convincing that shortly after the publication of Darwin's work, the vast majority of biologists had already accepted his theories of the descent and differentiation of species.

Observations of mockingbirds and turtles on the Galapagos Islands gave Darwin important clues for his **fourth thesis**. But the idea of a gradual change in organisms was alien to his contemporaries. The obvious idea that animal and plant breeding were key witnesses for Darwin's fourth thesis remained unconsidered. For Darwin's explanation for the variability of individual traits broke with everything that was sacred to science: new traits were supposed to arise *by chance* and then be transmitted to the offspring via an unknown mechanism. This idea seemed even more abstruse to biologists than to physicists, who had just seen chance break onto the scene at about the same time. The natural scientists of the nineteenth century were not willing to accept chance as a scientifically explanatory principle. It was neither compatible with their world of experience—successful breeding requires the consciously controlled action of a breeder—nor with the Christian principles of faith. Nor could they imagine any mechanisms on which such random variations could be based. Unlike the first three theses, the thesis of random variation—the core of Darwin's theory of evolution—never prevailed during his lifetime.

Darwin was also unable to provide solid evidence for his **fifth thesis**, the theory of natural selection. This was not least due to the fact that he had no scientifically validated explanation for how a change that arose by chance could be preserved through the generations. After all, it was assumed that an inherited trait would be mixed in equal proportions in the offspring; only half of a randomly evolved, advantageous trait of one parent would be passed on to the next generation, only a quarter to the one after that, and so on. According to this principle, variation would be "diluted" into oblivion in just a few generations; gradual selection on the basis of changed traits could not therefore take place.

In his 1868 work "The Variation of Animals and Plants under Domestication", Darwin presented a hypothetical path for heredity that would solve this problem: every part of a living body should continuously emit small organic particles, the gemmules, which accumulate in the gonads and provide them with heritable information. But this theory of pangenesis, as he called it, was too speculative to convince his critics. Nor did it solve the problem of "dilution".

Why did Darwin fail to prove his theories of selection and variation? Kepler's planetary theory, Galileo's law of falling bodies, and Newton's law of gravitation were quickly accepted by their peers because the movements of the planets, and the effects of the forces and gravitation can be observed and understood any time we choose. Darwin's theory of evolution, on the other hand, deals with a process that eludes direct human observation:

- Every change in an individual is unique and cannot—as the rules of science would require—be reproduced in an experiment at any time and in any place.
- The impact of individual changes on the characteristics of a species span many generations and thus periods of time that can hardly be comprehended by a single person.

Despite these limitations, Darwin emphasised in his 1868 book that the way he derived and justified his theories was very similar to Kepler's or Newton's approach:

Alone I use the same method that has been used in judging the ordinary phenomena of life and has been employed by the greatest naturalists. In the same way one arrives at the theory of the wave motion of light, and the assumption that the earth moves on its own axis has until recently hardly been supported by direct evidence.

But no amount of clever rationalisation could help: Darwin's theories of gradual variation and natural selection were shared only by a few biologists of his time. His first three theories also caused Darwin problems. For the readers of the "Origin of Species," it was obvious to ask whether we humans might not also belong to a species that constantly evolves over time and possesses a family tree with disagreeable ancestors lurking somewhere near the beginning.

Man Becomes Part of the Animal Kingdom

In order to offer as little room for attack as possible, Darwin carefully excluded the delicate question of the descent of man in his work of 1859. Only at the very end of his book is there a brief mention of this subject:

> In the distant future I see open fields for far more important researches. Light will be thrown on the origin of man and his history.

As luck would have it, in 1856 construction workers in the **Neander Valley** near Düsseldorf discovered for the first time a human-like skeleton with a peculiar skull and other unusual features. The skeleton was recognised as "primeval". In the light of Darwin's theory, this find made sense: it might be a human ancestor! Strangely enough, Darwin, who was obsessed with detail, only mentioned the Neanderthal fossil in passing in his book published fifteen years later.

Despite this hesitant statement, Darwin was accused of saying that humans were somehow close relatives of apes. And he had to endure far more serious hostility than the caricatures appearing in English newspapers showing him as an ape. The never-ending controversy forced Darwin to take a clearer position on the question of human descent, but as before, he took his time to react. It was not until 1871, twelve years after his book on the origin of species, that he published his second major work, entitled "The Descent of Man, and Selection in Relation to Sex". In this book he formulated three decisive theses:

1. The principles of species evolution also apply to humans. The anatomical similarities with today's apes indicate that humans and apes have common ancestors.

2. Darwin defined sexual selection as a mechanism of central importance for the course of evolution via variation and selection. Only through the fusion of the genetic material of males and females could new individuals with unique characteristics in terms of colour, shape, expression, and function (variation) emerge again and again. Sexual selection is also decisive for selection. For adaptation to changing living conditions only ensures the survival of an individual; if it does not create any offspring, its evolutionary significance is of the same importance as that of an individual that dies early. Only when individuals are successful with the opposite sex do they pass on their traits and thus also play a role in the survival of their species. This is because the more different traits are developed within a population, the higher the chance that some of the individuals will be able to adapt to unfamiliar living conditions. Sexual reproduction is therefore of central importance for speciation and diversity in nature.

3. Darwin combined his two new theses of the continuous evolution of man and the importance of sexual selection and came to an astonishing realisation: it is not only the physical characteristics of a human being that have a positive effect on his or her ability to survive and reproduce—mental, moral, and social abilities are also important. Darwin had studied countless forms of behaviour, and in particular sexual behaviour, but also facial expressions, instincts, and social structures, and he compared them with the corresponding characteristics in the animal world. On the basis of his findings, he suspected that sexual selection was the reason for the development of man's enormous intellectual capabilities. The abilities required to create music, art, or literature hardly play a role in the survivability of an individual. But in sexual selection, such abilities matter. So, the display of morality, wit, intelligence, or charm has the same effect as beautiful bird calls, colourful petals, colourful butterflies, and iridescent peacock feathers in the animal and plant kingdoms. Darwin even suggested that human intelligence and imagination might have evolved *primarily* because of sexual competition.

Darwin and Boltzmann

A great follower of Charles Darwin's teachings was the physicist Ludwig Boltzmann. The nineteenth century should probably be called Darwin's century, he said, because he had developed "the most wonderful mechanical theory in the field of biological science". Boltzmann even made scientific contributions to Darwin's theory of evolution. For example, he tried to place the physical meaning of photosynthesis on a Darwinian basis. Boltzmann described the general life processes from the point of view of his field as follows:

> The general struggle for existence of living beings is therefore not a struggle for basic substances – the basic substances of all living beings are abundantly available in air, water, and earth – nor for energy, which is abundantly contained in every body in the form of heat, unfortunately unconvertible, but a struggle for entropy, which becomes disposable through the transition of energy from the hot sun to the cold earth. To make the most of this transition, the plants spread out the immense surface of their leaves and force the solar energy, in a still unexplored way, before it sinks to the temperature level of the earth's surface, to carry out chemical syntheses of which we still have no idea in our laboratories. The products of this chemical kitchen form the object of struggle for the animal world.

Boltzmann and Darwin, already outwardly similar with their magnificent beards, both introduced random processes and the principle of irreversibility into their respective subjects—only Darwin dealt with populations and Boltzmann with physical systems.

- Darwin showed that the emergence of variations in a population is determined by random, uncontrolled changes.
- Boltzmann showed that even the ordered laws of physics contain a dose of random, uncontrolled events, and that statistical calculations are unavoidable if one wants to understand the world.

One of them shook up the traditional creation story and the creationism of strict Christians with his theory of evolution, the other overthrew Newton's mechanistic universe with his entropy. But the triumphal march of the respective new world views was quickly thwarted. In physics, the focus was on the smallest particles, whose properties were still unknown. And in biology, it was still unknown how properties were transmitted from one generation to the next.

Darwin had no idea that in the garden of an Austrian monastery near Vienna, work was already being carried out tirelessly on a link between the theory of evolution and the physics of chance.

The Atoms of Life

Gregor Mendel was a monk and later also abbot of the Augustinian Order in Brno (today the Czech Republic). For eight years from 1856—during which time Darwin published his Origin of Species and Boltzmann was about to introduce statistics into thermodynamics—Mendel was working on his now world-famous crossbreeding experiments. In addition to his theological training, Mendel had also had a sound scientific education: he had studied physics in Vienna from 1851 to 1853 and had learned, among other things, how to systematically evaluate experiments on the basis of the kinetic theory of gases. Mendel thus carried out his crossbreeding experiments with clear theoretical ideas.

While Darwin only paid attention to the *changes* in traits over the generations, Mendel focused on the *persistence* of their transmission. He wanted to find out how plants pass on their characteristics to their descendants and why some seem to disappear and only reappear after a few generations. He chose the pea as the object of his experiments. Its large flowers allowed him to precisely transfer pollen from one plant to another. In addition, a large number of pure-bred varieties were available with easily distinguishable characteristics. Mendel selected pea varieties that clearly differed in seven characteristics: high and low growing, green and yellow, smooth and wrinkled seeds, white and purple flowers, etc. In his crossbreeding experiments over several generations, he meticulously recorded the characteristics of a total of about 28,000 plant individuals. Thanks to this literal bean counting, Mendel made three important discoveries:

1. **Uniformity rule**: The progeny of two purebred plants of the same variety crossed with each other have the same characteristics as their parents.
2. **Cleavage rule**: If two different varieties are crossed, the trait of one parent is retained in its entirety, while the trait of the other parent is not expressed. Some traits appear reliably in the daughter generation, and these Mendel called "dominant"; others seem to disappear in the succeeding generation but reappear in later generations, and these he called "recessive". Mendel discovered that the expression of dominant and recessive traits in later generations is random and thus unpredictable in individual cases, but in the longer term follows clear mathematical rules. Just as the probability of rolling a certain number of dice is 1:6 after a sufficiently frequent roll of the dice, a sufficiently high number of individuals produced by crossing different varieties results in a ratio of dominant to recessive traits equal to 3:1.

3. **Independence rule**: Different traits such as pea size and colour are inherited independently of each other. For example, if large, yellow peas are crossed with small, green peas, only large, yellow peas will appear in the first generation, because both characteristics are dominant. In the second generation, the characteristics then combine independently of each other. Subsequently, there are large green, large yellow, small green, and small yellow peas.

The **ratio of 3:1** dominant to recessive trait carriers in the second generation is based on an internal ratio of 1:1:2. One part of the plants is purely dominant, another purely recessive, two parts are mixed dominant–recessive. Since in their case the dominant character prevails externally, the appearance of the traits results in a ratio of 3:1.

Today we know that Mendel happened to have selected the "right" traits for his crossbreeding experiments with peas. This is because his rules only apply if the genes responsible for the traits in question are located on different chromosomes or are sufficiently far away from each other on the same chromosome.

Darwin had already introduced *chance* into biology; thanks to Mendel, *statistical relationships* now also found their way into biology for the first time. This was another astonishing parallel with almost simultaneous developments in physics.

A more precise examination of Mendel's data by the statistician and population geneticist Ronald Fisher some sixty years later showed that Mendel's results fitted his theory far better than could be expected statistically. Because the ratio of dominant to recessive phenotypes was statistically too close to the ratio of 3:1, it was suspected that Mendel embellished his results when he realised that they were heading towards this ratio. However, Fisher's analysis is also controversial today. The Fisher–Mendel controversy continues to preoccupy statisticians to this day.

Approach to the Gene

Because even a tiny fertilised egg cell contains all the information needed to give rise to a whale, an oak tree, or any other living creature, it was agreed that the potential for inheriting certain characteristics must be stored somewhere in the tiniest, as yet unknown cell structures of living beings—and thus lay beyond all hope of observation at the time. The laws discovered by Mendel, however, allowed conclusions to be drawn about further properties of heredity:

- Inherited traits are not "miscible"; a tall pea plant crossed with a low-growing pea plant will not produce medium-sized individuals.
- Individuals can pass on traits to offspring even if they are not expressed in themselves. They travel through the generations as "stowaways" and can reappear at some point.

Everything pointed to the fact that, in heredity, *indivisible* smallest units act as carriers of hereditary information. These might be called "atoms of heredity".

Unfortunately, Mendel's achievement was not recognised during his own lifetime and his findings only found their way into the scientific discussion after a long delay. He had carried out his experiments in monastic solitude and published his essay "Versuche über Pflanzen-Hybriden" (Experiments on Plant Hybrids) in 1866, in a small volume edited by a local naturalist association that he himself had co-founded. He sent copies of the volume to the leading biologists of the time, but they obviously did not like mathematical style of the publication. Thus, Mendel's discoveries remained largely unknown to the scientists of the nineteenth century.

Darwin had brought variation and selection into play with his theories, without, however, being able to explain how a variation that arose by chance could be preserved until selection came about, simply because selection requires far longer than the lifespan of an individual. Because it was assumed that a trait that arose by chance would be "diluted" from generation to generation until it soon disappeared, the link between variation and selection was still missing. It was precisely this *missing link* between random variation and selection determined by the environment that Darwin had searched for in vain.

Anyone who knew the work of both researchers would have been able to draw the decisive conclusion that tiny random changes in genetic material can indeed result in clear macroscopic effects. Did Charles Darwin know the work of Gregor Mendel? It is not unlikely that Mendel had also provided Darwin, who was already very well known at the time, with a copy of his work. But Darwin in particular would probably have been put off by the statistical calculations, for he was of the opinion that "mathematics is as useful in biology as a dissecting knife is to a carpenter". Presumably, however, Darwin knew of Mendel's work indirectly: in at least two books that he can be shown to have read, Mendel is quoted at length; and one of them has Darwin's handwritten notes in close proximity to this passage. Even if Darwin did read Mendel's findings, he obviously did not recognise the connection with his own theory of heredity.

It was not until a generation of researchers later, in 1900, that Mendel's work was rediscovered by the Dutch biologist Hugo de Vries. He established the connection between Darwin's theory and Mendel's observations and thus heralded the birth of genetics, and with it, modern biology. In 1901, in his work "The Mutation Theory", he introduced the term mutation into the theory of heredity (from the Latin word *mutare*, "to change"). Until then it had only been used in palaeontology. De Vries was also one of the first to advance Darwin's hypothesis of particular hereditary carriers. As early as 1889, in his book "Intracellular Pangenesis", he had postulated that the various biological characteristics were inscribed on individual hereditary carriers. De Vries called these units "pangenes" after Darwin's theory of pangenesis, and spoke of "genetic mutations".

The term "gene" (from the Greek *genos* "offspring, kinship" or *genesis* for "creation" or "development") was first used in its current form in 1909 by the Danish botanist Wilhelm Johannsen. The gene was supposed to contain all the necessary information about the characteristics of an individual that could be inherited. Thus, forty years after Mendel had postulated particle-like inheritance elements and Darwin germ cell inheritance carriers, biologists gave the "atom of biology" a name. Like the atom in physics, the gene now became the central concept in biology.

New Theories Without Answers

By 1900, biologists had long outgrown the task of observing and describing animals and plants and classifying them in taxonomic systems. In practice, thanks to their insights, plant breeders were able to celebrate great successes, but in theory their discipline contained fundamental gaps. At that time, biology—like physics and mathematics—was in need of further explanation.

- The adventurer Darwin had explained the development of life on Earth on the basis of natural principles. The world as God's creation was no longer a valid explanation for scientists. But there was no alternative interpretation of the processes of life. The purely empirical laws of evolution and rules of heredity could not even be given a rudimentary explanation; an internally consistent foundation was still missing.
- Mendel, the bean counter, had for the first time formulated concrete laws of heredity with the help of statistical correlations and thus brought the unloved laws of chance (probability) into play.
- For a certain time, the gene theory remained a mere hypothesis that only raised more questions: Where were they, after all? What were they made of? And how did they control heredity? Just as physicists could not observe atoms, biologists were unable to study genes directly. Therefore, the existence of genes seemed very questionable to many scientists of the late nineteenth and early twentieth centuries.

The detailed mechanisms and functional principles of life seemed to be hidden in the cells of plants and animals, inaccessible to observation with a microscope and thus undetectable. In physics as in biology, it was true that the smallest particles determined everything, but no one had any idea how that could work.

5

The Ground Slips from Under Our Feet—The Collapse of the Classical Sciences

The second half of the nineteenth century was a golden age for the natural sciences. Scientists accumulated fundamental knowledge about natural events at breakneck speed, so that physics had an almost complete and all-explanatory body of theory at its disposal:

- Newton's theory provided the foundations of mechanics.
- Dalton's atomic theory, which assumed solid, small atomic spheres, made the structure of the material world comprehensible.
- Boltzmann's kinetic theory of gases, based on Dalton's atomic theory, made it possible to understand thermodynamic processes.
- Maxwell's field theory explained the phenomena of electricity, magnetism, and optics.
- Darwin's theory of evolution and Mendel's genetics had tracked down the laws of heredity.

The essential questions about how the world works were thus considered answered. Even if part of the scientific world had not kept up with these innovations, it was generally believed that the "few details" that did not yet fit into the picture would soon fit into the existing theories. When the young Max Planck asked his professor at the University of Munich, Philipp von Jolly, in the 1870s whether he should study physics, he was advised against it on the grounds that "there is not much left to discover in this field".

© The Author(s), under exclusive license to Springer Nature
Switzerland AG 2022
L. Jaeger, *The Stumbling Progress of 20th Century Science*,
https://doi.org/10.1007/978-3-031-09618-1_5

Outside the scientific world, too, the mood was euphoric, as the successes of the researchers were translated into technological achievements and their possibilities for improving the world seemed limitless. Ever more efficient steam engines, increasing mechanisation, and the triumph of electricity triggered a second industrial revolution.

- Electric light illuminated houses and streets at night; today it is hard to imagine the liberation from gloom and confinement this meant for people.
- Thanks to the telegraph and the telephone, news could be exchanged on a daily basis with the most distant parts of the world.
- People and goods were transported with unprecedented speed, while international trade increased by leaps and bounds. Today we speak of the "first globalisation".
- Life became noticeably more comfortable thanks to technologies such as central heating, trams, ways of preserving foods and, last but not least, the automobile.
- Extraordinary breeding successes and the use of artificial fertilisers could now ensure the food security of many people.
- Researchers like Rudolf Virchow, Louis Pasteur, and Robert Koch discovered how diseases develop and thus found ways to treat them. Diseases like cholera lost their terror.

The result was the most massive and comprehensive increase in quality of life and prosperity ever witnessed throughout human history. Especially in Europe and North America, people's everyday lives and perceptions changed faster and more profoundly than ever before. Cities like Berlin, London, Paris, New York, and Chicago were vibrant with optimism and vitality. No longer were politics, the behaviour of rulers, and wars seen as the driving forces of all social development, but scientific and technological progress.

The successes of the natural sciences led to their principles and methods also becoming popular in other fields of knowledge. Thus, in the social sciences, the view spread that the development of human society must obey laws similar to those in nature. The most influential social and economic theory of the nineteenth century, the economic theory of Karl Marx and Friedrich Engels, applied the mechanistic-deterministic principles of physics to social change and concluded that a lawful development of society towards communism was as clearly predetermined as the movement of the planets. Other social scientists took Darwin's theory of natural selection, applied it to human society, and used it to explain the rationality and legitimacy of the inhuman system of early capitalism.

A Short Excursion into Philosophy

The successes of scientific theories and the power of the technologies resulting from them seduced many people into believing that there could be no suitable explanations for what happens in the world outside the natural sciences. The name for this current of thought in philosophy is positivism. According to this explanation of the world, only direct experience and what can be inferred from it by reason and logic can lead to real knowledge. The most prominent representative of positivism was Ernst Mach, whom we already met in the second chapter. In the philosophical debate between idealism, rationalism, and empiricism, positivism occupies a position of its own.

- For **idealism**, which was widespread in Germany during the early nineteenth century, our perceptions and knowledge arise not so much from our external experiences and our minds, but primarily from the *ideas* that exist within us. Thus, the physical world actually exists only as an object in our consciousness. The positivists set themselves apart from this current of thought with their demand that the world should be grasped with as much economy of thought as possible: only what is directly perceived is relevant, while vague statements and metaphysical assertions have no meaning.
- **Rationalism** assumes that the *human mind* is crucial for the way we process our experiences and thus recognise reality and truth; the experiences themselves play a minor role. Descartes' famous phrase "I think, therefore I am" sums up this philosophy. Positivists have fairly similar attitude in some respects, but the fact that in rationalism even morality, ethics, and the social order are justified by reason goes too far for them.
- According to **empiricism**, all knowledge is based exclusively on *sense experience*. The fact that this world view entails a profound problem had already been grasped by the pre-Socratic philosopher Heraclitus. "No one gets into the same river twice." Since sensory experience is different at each point in time, no universally valid knowledge can be obtained from it. This view was echoed by Gottfried Wilhelm Leibniz in the seventeenth century, when he pointed out that every sensory experience is always an isolated case. David Hume took this idea further in the eighteenth century: according to his so-called problem of induction, a generally valid law endowed with absolute certainty can never emerge from an individual case. Empiricism also has a problem from a logical point of view: it cannot satisfy its own principles, because propositions such as "All experiential knowledge is true" or "Valid knowledge is based solely on sense experience" cannot be derived

from experience. The positivists tried to free themselves from the philo-sophical problems of empiricism by changing the goal of their search: it is not the claim to *truth* that is decisive for the relevance of a theory, but solely its *usefulness* in describing and predicting phenomena.

Positivism: A widespread attitude among natural scientists at the end of the nineteenth century, which assumes that reliable knowledge can only be based on the interpretation of "positives", i.e., on findings that are perceived by the senses and empirically verifiable.

In positivism, the best theory is the one that best describes and predicts our sensory impressions. Should experience and data later require a different theory, another one that can better capture the observations will take its place. This world view gives rise to a new and important epistemological maxim: it is not only life, but also scientific theories that are subject to evolution.

The Fundamental Limits of Scientific Knowledge

Optimism about progress dominated in those days. But there was also an undercurrent, initially slight and later gaining strength, that pointed in a completely different direction. In August 1872, at the Assembly of German Naturalists and Physicians in Leipzig, the generally widespread positivism was called into question. The physiologist Emil Du Bois-Reymond outlined his credo *Ignoramus et ignorabimus*, meaning: We do not know *and* will never know, in his speech "On the Limits of the Knowledge of Nature". His lecture ended with the words:

> In the face of the riddles of the physical world, the natural scientist has long been accustomed to pronounce his "Ignoramus" with manly renunciation. Looking back on the victorious path he has travelled, he is carried along by the quiet consciousness that where he does not know now, he could at least know under certain circumstances, and perhaps will know one day. But in the face of the riddle of what matter and force are, and how they are able to think, he must decide once and for all on the much more difficult truth: 'Ignorabimus'.

What a slap in the face for those present! There was general agreement that we were on the verge of a final explanation of *all* phenomena and experiences and that a complete and self-contained, scientifically based world view was within our grasp! It was clear to the audience: if the insurmountable barriers to scientific knowledge described by Du Bois-Reymond really existed, this would mean that a scientifically founded world view would forever have gaps that could only be filled with idealistic, religious, and mystical elements, but never with rational or positivistic explanations.

Du Bois-Reymond was secretary of the Prussian Academy of Sciences and one of the most influential representatives of the German natural sciences—his view could not be ignored. It was rejected by many scientists, but thanks to the persuasiveness of his arguments there were enough supporters to ensure that it was discussed again and again. Despite the opposition, Du Bois-Reymond remained true to his view over the following years. In another speech in 1880, entitled "The Seven World Riddles", he gave specific form to his sceptical stance and named the riddles he considered could never be answered:

1. What are matter and force?
2. What is the origin of motion?
3. How did life first come into being?
4. Where does the purpose ("the deliberate purposeful arrangement") of nature come from?
5. How does consciousness ("a conscious sensation") arise from unconscious nerves?
6. Where do rational thought and "closely related" language come from?
7. Where does human free will come from?

These are precisely the questions that have been debated since antiquity and in almost all cultures. The fact that they have lost none of their mystery to this day strengthens Du Bois-Reymond's position, although it was hotly contested at the time.

Questions 5 and 7 in particular are currently under intensive investigation. Despite our greatest efforts, we are still far from grasping the essence of free will mentioned in problem 7—today also referred to as "*intentional* consciousness". Science has got somewhat further in explaining conscious sensation—neuro-philosophers today speak of "*phenomenal* consciousness" or "qualia". Some possible explanations have been found, and many a scientist hopes to solve the puzzle known in today's philosophy as the "qualia problem", but basically we have not progressed much further here than Du Bois-Reymond had already summarised in his lecture of 1872:

What conceivable connection exists between certain movements of certain atoms in my brain on the one hand, and on the other hand the facts that are original to me, that cannot be further defined, that cannot be denied: 'I feel pain, I feel pleasure; I taste sweetness, smell the scent of roses, hear the sound of an organ, see red …' and the certainty that also flows from this: 'Therefore I am'. From this it is quite and forever incomprehensible that a number of carbon, hydrogen, nitrogen, oxygen, etc. atoms should not be indifferent to each other. Atoms should be indifferent to how they lie and move, how they lay and moved, how they will lie and move. It is utterly inconceivable how consciousness could arise simply from their being together.

Du Bois-Reymond was anything but an opponent of science. In his field, physiology, he advocated a consistent research programme, for he was convinced that natural science could, within certain limits, achieve considerable gains in knowledge; for him, the history of natural science was "the real history of humanity". And yet, among those who were not dazzled by the successes of science, he was the first to publicly question its omnipotence with his *ignorabimus*. Others despaired on a personal level that certain problems in the accepted theories of physics and mathematics refused to disappear. Boltzmann, who found no solution to the problem that his entropy formula failed in Gibb's two-bin experiment (see Chap. 2, page 27/28), and Cantor, who became entangled in the contradictions of his infinities, were both seriously depressed. We can only speculate on how far their harrowing experience that the worlds of science and mathematics stood on feet of clay contributed to Boltzmann's suicide and Cantor's tragic end in an insane asylum.

Ignorabimus in Mathematics

Positivism had never really been able to gain a foothold in mathematics. Leibniz had already recognised that there is no sensory experience in this discipline, above a certain level of abstraction and the fact that mathematical truths must therefore come from somewhere else. When Du Bois-Reymond made his scepticism public, the existential debate was taking place in mathematics about how to think about the infinite without contradictions. The problems of the mathematicians did not remain hidden from scientists working in other disciplines. Even biologists took part in the discussions. One of the founders of microbiology, Louis Pasteur, stated:

> The great and obvious gap in the system is that the positivist worldview takes no account of the most important of positive ideas, the idea of the infinite.[1]

---

[1] Letter to his son-in-law Jean Martel, who also served as his secretary, published in 1881 and initially anonymously, M. Pasteur, *Histoire d'un savant par un ignorant*. Large parts of this book were written under Pasteur's direct supervision, which means that this work could also be considered an unofficial "autobiography".

However, the most influential mathematician of his time, David Hilbert, found the idea that human knowledge should be limited too pessimistic. His speech to the Paris International Congress of Mathematicians in 1900, mentioned in the third chapter, should be seen against the background of the controversy surrounding Du Bois-Reymond's *Ignorabimus*. Hilbert insisted on his opinion that answers to *all* problems can be found in mathematics:

> This conviction of the solvability of every mathematical problem is a powerful incentive for us during our work; we hear the constant call within us: 'There is the problem, look for the solution! You can find it by pure thinking! For in mathematics there is no ignorabimus.

The fact that Cantor struggled with the "real" infinities was of no fundamental importance to Hilbert. For in Hilbert's view, since these infinities are neither real in nature nor logically conceivable in our thinking, they had to be theoretical exercises with no effect on the foundations of mathematics. He published this opinion as late as 1930:

> We see that the infinite is nowhere realised; it is neither present in nature nor admissible as a basis in our thinking without special precautions. In this alone I see an important parallelism between nature and thought, a fundamental agreement between experience and theory.[2]

Hilbert also rejected ignorabimus for physics, because in his view mathematics stood at the heart of every natural science:

> We do not master a scientific theory until we have peeled out and completely revealed its mathematical core. Without mathematics, contemporary astronomy and physics are impossible; these sciences virtually dissolve into mathematics in their theoretical parts.[3]

But the positivists were gradually losing more and more ground. The more they saw themselves cornered, the sharper and also the more desperate the tone became. Hilbert again:

> Anyone who feels the truth of the generous way of thinking and world views [...] does not fall into regressive and unfruitful doubtfulness; he will not believe

---

[2] D. Hilbert, *Naturerkennen und Logik*. Naturwissenschaften 1930, pp. 959-963 (also published in: Gesammelte Abhandlungen, vol. 3, p. 378; available at: https://www.rschr.de/Htm/David_Hilbert_Naturerkennen_und_Logik.htm.

[3] D. Hilbert, *Naturerkennen und Logik* (1930), p. 385 (also published in: Gesammelte Abhandlungen, vol. 3, p. 378; available at: https://www.rschr.de/Htm/David_Hilbert_Naturerkennen_und_Logik.htm.

those who today prophesy the end of culture with a philosophical bearing and a superior tone and take pleasure in ignorance. For the mathematician there is no ignorance, and in my opinion, there is none whatsoever in natural science. Instead of foolish ignorance, our slogan should be: "We must know - we will know."[4]

With these words, the belief in the possibility of complete knowledge in mathematics manifested itself for the last time. Just a year later, Kurt Gödel proved his incompleteness theorems and thus anchored *ignorabimus* in mathematics forever. His proof rules out any possibility of getting rid of it at some later time, no matter what mathematical systems become accessible to us as we gain further knowledge.

No Longer Master in One's Own House

It was a great challenge for scientists to get used to the idea that their theories were by no means as clear and accurate as they had assumed. Too many open questions disturbed the "scientific peace".

- Boltzmann and the inexplicable statistical behaviour of atoms.
- Cantor and the recalcitrant infinite sets.
- Darwin and the mysterious origin of life.
- Mendel and the inexplicable nature of genes.
- Planck and quantised radiation.
- Einstein and the interwoven structure of space and time.

Insidiously, all these inconsistencies had come to threaten the theories of the different scientific disciplines. Most of the scientists mentioned had been expecting to take care of the "last little problems," but instead they had added theories that did not just putty over the cracks that had come to light, but only widened them. Einstein's theory of relativity and his quantum theory proved to be genuine wrecking balls that completely toppled the already tottering construct.

Many researchers had thought that in the foreseeable future they would be the all-powerful masters of the world, but now they were forced into the role of subalterns who would have to content themselves with wresting one or two last secrets from nature. It was precisely in this highly unstable

---

[4] ibid.

context that a study was published that ushered in another upheaval in self-perception. In the field of psychology, all the conscious aspects of the human psyche had been studied and it was assumed that the essence of the human mind had thereby been grasped. But the Austrian physician Sigmund Freud, in his book "The Interpretation of Dreams" published in 1899, completely recharted human consciousness:

- In addition to the conscious part of our psyche that we can grasp and reflect upon, there is also an unconscious part that we do not directly recognise or experience.
- Man is not free in his feelings, thoughts, and actions; rather, every action or utterance, no matter how insignificant, is decisively influenced and controlled by his uncontrollable subconscious.
- The cause of many mental illnesses lies in the repression of these unconscious thoughts, desires, and emotions.

The distinction between the conscious and the unconscious was not entirely new; for example, the writings of the Russian writer Dostoevsky, whom Freud deeply admired, contain numerous passages describing an unconscious part of our psyche. But the fact that Freud assigned the unconscious the dominant role and used methods such as hypnosis, which according to popular opinion belonged to the realm of quackery, aroused great displeasure among many natural scientists of his time. The theories according to which the causes of many psychological disorders lie in a repressed but ever present desire for sex, or that sons at a certain stage of development compete with their father for their mother's attention—the Oedipus complex—also met with widespread rejection in the age of outward conservativism that prevailed at the turn of the century. This went so far that his colleagues refused him access to laboratories and clinics where he wanted to test his methods. It was not until the First World War and the return of the many soldiers suffering from severe psychological disorders that Freudian psychoanalysis finally achieved a breakthrough.

Freud himself described the discovery of the unconscious and its central role in our lives as a further "degradation of man". After Copernicus, man could no longer see himself at the centre of the universe; after Darwin, he was no longer the crown of creation; and now with the theory of the unconscious he was no longer even "master in his own home".

Some did not know how to proceed, but others like Niels Bohr and Einstein were less pessimistic and highly motivated to actively pursue the contradictions, for they knew they were on the trail of something completely new.

Early Philosophical Crises: Nihilism and Existentialism

Outside the sciences, too, optimism about progress was supplanted by other moods from the first years of the twentieth century onwards. While the progress and optimism of the nineteenth century had shouted "Keep it up!" to the world, there was now a strong desire to break new ground, to break away from the old. Many people sensed intuitively that major shifts were on the horizon. For some, who would have been only too happy to continue to make themselves comfortable with what had already been achieved, this meant a further loss, while others could hardly wait for the new era and found themselves roused into a mood of innovation and novelty.

In parallel with the upheavals in physics, mathematics, biology, and psychology, a sense of unease was spreading through the societies of Europe and America. Many people felt increasingly insecure in their values and self-perception and psychologically overwhelmed:

- As mechanisation and industrialisation progressed, people became increasingly alienated from themselves, from the faith that could have sustained many, and from nature.
- The complexity of the world meant that it could only be grasped in fragments.
- The new speed of life was perceived as frantic; many "could no longer keep up".

The general unease and the conscious move away from tradition also found expression in art, architecture, music, and literature.

- In the visual arts, avant-garde movements such as Expressionism and Cubism came on the scene, and many young artists tried their hand at abstraction, developing it to the point of non-objectivity.
- In architecture and interior design, it became fashionable to remove all superfluous decoration—Art Nouveau was just one of the many expressions of the new longing for clear forms and lines.

- From around 1910, "new music" deliberately broke with the rules of classical music; among others, the Viennese composer Arnold Schönberg developed the twelve-tone technique.
- In literature, the sensation of a general instability came to the fore. Examples are the works of Herbert George Wells, Hugo von Hofmannsthal, Rainer Maria Rilke, Franz Kafka, Oscar Wilde, and the brothers Thomas and Heinrich Mann.

Probably the most significant manifestation of this extreme pessimism occurred in philosophy: In the last twenty years of the nineteenth century, a mindset known as nihilism came into being. A nihilist believes in nothing, and has no loyalties and no goal, except perhaps the urge to destroy. Nihilism is most often associated with Friedrich Nietzsche, who argued that the corrosive effects of philosophical thought would ultimately destroy all moral, religious, and metaphysical beliefs, and bring about the greatest crisis in human history.

Nihilism: According to this worldview, all knowledge and values are basically worthless. Nothing can really be grasped or communicated. This was associated with a radical scepticism that rejected even the real existence of many things.

All these developments can be interpreted as cultural pessimism, an end-of-time mood, also referred to as the "fin-de-siècle" attitude to life, and a reaction to the spiritual–cultural upheaval and collateral damage of industrialisation.

Thus, it came about that the turn of the century was simultaneously characterised by a strong mood of optimism and euphoria, as well as a diffuse fear of the future and an end-of-time mood. There was a strange mixture of optimism and pessimism, and the shearing forces to which the individual was exposed were enormous. Then, the "primordial catastrophe" of the twentieth century, the First World War, burst upon the highly unstable state of Western society.

Total War

Since 1870/71 when the last great war was fought in Europe, technological developments had led to an arms race between the European powers and produced a military potential whose destructive power was simply unimaginable to humankind. In addition to developments in transport and communications, there had been advances in the industrial production of ammunition and ballistics; the scientific basis for sound detection and the production of poisonous gases had been established, and medical care was also a factor in warfare, as it massively tied up forces by dealing with the wounded.

The assumption often made before 1914 that a modern war would have to be more humane because of advanced weapons technology proved to be fundamentally wrong. For many soldiers on both sides, the war had begun with the expectation of a fresh and quick victory. But the effect of the new technologies and the brutality of their use shook them to the core.

- **Machine guns**: The common tactic in the field was initially the same as in the previous war of 1870/71: foot troops with rifles and cavalry brigades made frontal attacks on enemy positions. But in this war, the soldiers were mown down by machine guns. In the first border battles at the end of August 1914, around 40,000 soldiers died within five days on the French side alone—mostly from explosives and machine guns. Their deaths were practically worthless in military terms. This was also the case with the 260,000 German soldiers who died or were wounded in September 1914, representing almost 17% of the forces deployed. The romanticism of an infantry or cavalry charge was quickly dispelled.
- **Tanks**: The American Benjamin Holt actually wanted to drive tractors and construction machinery with tracked vehicles (his company later became the world's largest bulldozer manufacturer "Caterpillar"). In 1904, the first functional tracked vehicle was ready for series production. Despite initial problems during the war, the British Secretary of the Navy, Winston Churchill, was convinced that such "landships" could overcome the German trenches. For reasons of secrecy, the English called these "water carriers." This gave rise to the name "tank", which is still common in English-speaking countries today. Tanks played an important part in the success of the Allied offensive in the early summer of 1918.
- **Aircraft**: At the beginning of the war, planes were mainly used for reconnaissance. In April 1915, the French pilot Roland Garros used an aircraft armed with a rigidly mounted, forward-firing machine gun in aerial combat against the German Albatros biplanes. This first fighter plane in history also inspired the German army command to build fighter planes. By the end of the war in 1918, the German air fleet had grown from an initial 300 aircraft to 5000. The Allies had even more aircraft at their disposal; by the summer of 1917 they had achieved air supremacy through numerical superiority. Overall, the military importance of aircraft for the

war effort was rather slight until 1918, but people were terrified by the prospect of war in the air.

- **Poison gas:** One of the technological inventions with the greatest impact on the history of the war was a process for the synthetic production of ammonia. Five years before the outbreak of the war, Fritz Haber and Carl Bosch had made the industrial production of fertilisers possible thanks to the process named after them. This was and still is the only way to defeat hunger in large parts of the world. But here, too, the ambivalence of scientific success is evident: the ammonia produced by the Haber–Bosch process was used to manufacture enormous quantities of explosives that cost the lives of millions of people. Fritz Haber was also directly involved in the production of weapons of mass destruction. Under his direction, chlorine gas was used for the first time in 1915, and later the even more toxic phosgene was added to the poison gas. His processes for the production of ammonia had saved millions and billions of people from starvation; at the same time, as the "father of the gas war", he was responsible for the cruel death or long suffering of thousands of people. Haber's wife Clara, also a chemist, described her husband's contribution to warfare as a "perversion of science". When Fritz Haber was promoted to captain after the use of poison gas in the Battle of Flanders, she shot herself with his service weapon.[5]

All in all, the almost eleven million dead civilians as well as the almost ten million dead and 21 million injured soldiers—not to mention those suffering from psychological trauma—constituted a massive collective shock. The old order had collapsed, and politics, society, economy, and culture had been completely transformed by the war. **New values and societal norms** led to a genuine rupture of civilisation, which caused many to feel completely disorientated.

---

[5] A comprehensive account of the history of chemical warfare agents in the First World War, which is less well-known in public discussion, is Ludwig (Lutz) F. Haber, *The Poisonous Cloud. Chemical Warfare in the First World War*, Oxford 1986. The author is Fritz Haber's son. See also F. Schmaltz, *Kampfstoff-Forschung im Nationalsozialismus*, Göttingen (2005), and the biography of Fritz Haber by Margit Szöllösi-Janze: *Fritz Haber 1868–1934*, Beck, Munich (1998).

From International Cooperation to National Polemics

The sciences were also dramatically disrupted as a result of the war. Before 1914, the spirit of natural scientists and mathematicians was internationally collegial. Thus, at the Prussian Academy of Sciences, on the occasion of the emperor's birthday in 1878, Emil Du Bois-Reymond wrote:

> Science alone is, by its very nature, cosmopolitan [...] All cultural peoples participate in the expansion of science to the extent that they deserve this name.

The spirit of cooperation, friendship, and internationality came to an abrupt end with the start of the First World War. Scientists on both sides were mobilised or volunteered to lend their knowledge to their country's victory. Attempts by individuals to save the community went unheard, including the statement by nine English professors who warned against Britain entering the war against Germany on 1 August 1914, the day the war broke out, emphasising "Germany's leading role in art and science". The physicist Joseph John Thomson and the chemist William Ramsey, both Nobel Prize winners, were among this group.

On 4 October 1914, 93 German scientists, among them Max Planck and five other Nobel laureates, signed a polemical appeal "To the Cultural World". It was a reaction to the protests by English scientists and intellectuals after German troops had set fire to the venerable library of the University of Leuven on 25 August 1914 during their invasion of Belgium. In their appeal, Planck and his colleagues articulated their "protest against the lies and slanders" of "Germany's enemies" and spoke of a defensive war, an "imposed, difficult struggle for existence". In essence, they adopted the position of the German military. The reaction of the scientists on the Allied side was not long in coming: on 21 October 1914, 117 English scholars, now including Thomson, who had previously wanted to exert a moderating influence, drew up a counter-declaration. They declared the fight against "militaristic Germany" as necessary and the British entry into the war as a "defensive war, a war for freedom and peace".

Among the few scientists who were not infected by patriotic war enthusiasm and the formation of opposing camps was Albert Einstein. On 1 August 1914, he wrote to the Dutch physicist Hendrik Antoon Lorentz:

> If a group of people are labouring under a collective delusion, these people should be deprived of all influence; but hatred and bitterness cannot control

great and farsighted people in the long run, this will let them be sick themselves.

He wrote to his friend Paul Ehrenfest, Lorentz's successor in the old university town of Louvain:

Unbelievable things have now begun in Europe in its madness. At such times one sees what a sad breed of cows one belongs to.

But Einstein was an exception. The international scientific community was as divided during the war as European societies as a whole. Within just a few weeks, the world of formerly friendly scientists had disintegrated into hostile camps, while professions of internationality and cooperation seemed long forgotten. The scientific community was mobilised and militarised as never before. On both sides, scientists were drafted or volunteered to make themselves available to their country. The inventor of poison gas, Fritz Haber, summed up this patriotic spirit:

In war, the scholar, like everyone else, belongs to the fatherland, but in peace he belongs to humanity.

Not only did international cooperation and the personal integrity of some natural scientists suffer, the war also led to a very direct loss for the sciences. Historians assume that around 20% of Germany's young physicists died on the front line or as war invalids at home. The later famous mathematician Richard Courant, who fought in Belgium, was seriously injured and only narrowly escaped death. It is not unlikely that among the millions who died were scientists whose genius was comparable to Albert Einstein, Werner Heisenberg, or Wolfgang Pauli.

Today, the First World War is described as the decisive epochal threshold of modernity, resulting directly in subsequent events such as the October Revolution, Stalinism, Fascism, National Socialism, and finally the Second World War. What is less well known is just how much of a rupture this war meant for the scientific world. It had already been in a profound crisis at the outbreak of the war. The euphoric optimism of the second half of the nineteenth century had completely evaporated. The cracks in the theoretical edifices of all the major disciplines had led to great uncertainty among scientists; from the turn of the century onwards, there were growing doubts about the validity of scientific knowledge that had previously been regarded as fundamental, overthrowing the confidence that a self-contained system could be found to explain the world. Then the scientific community was shattered into hostile camps, followed by the terrible bloodletting of young scientists caused by the First World War. How could something new arise from these ashes? How could scientific progress continue?

Whatever Happened to the Quest for Truth?

In Germany, conditions were not conducive to a new beginning in the sciences. The lost war and the conditions of the Treaty of Versailles, which were perceived as humiliating, only strengthened nationalist feelings. The "stab in the back" myth—that the German armies had never been defeated, that only treachery on the home front had brought about their defeat— helped them to maintain their self-esteem. Since the Jews were held primarily responsible for the "stab in the back", strong anti-Semitic feelings were mixed into this nationalism, with serious consequences for science as well.

Albert Einstein's theory of relativity in particular was now sharply attacked by national conservative physicists. The spokesmen for this movement were the physicists Rudolf Tomaschek, Philipp Lenard, and Johannes Stark, the latter two having received the Nobel Prize for Physics in 1905 and 1919, respectively. They labelled Einstein's theory of relativity as "Jewish physics" and opposed it with "German physics". They ignored the fact that the theory of relativity had long been confirmed by a series of observations:

- The orbit of Mercury shifts by 10,000 kms every hundred years. In astronomy, this is a tiny deviation, but only Einstein's theory of relativity could explain this irregularity.
- The Michelson–Morley experiment showed that the speed of light remains the same even if it is measured once in the same direction as the Earth moving around the Sun and once in the opposite direction.
- The deflection of the Sun's light during a solar eclipse in May 1919 exactly matched the prediction made using relativity theory.

Stark, in particular, relied more on ideological than scientific principles, and thus tried to establish himself as the national authority on "German physics". However, Lenard and Stark's support from the NSDAP was not as great as they would have liked. The Nazis put their faith in the quantum physicist Werner Heisenberg, whom Lenard and Stark's circles promptly labelled a "white Jew". After a complete "character assessment" by the Nazis, Himmler forbade further attacks on Heisenberg. It certainly helped here that Heisenberg's grandmother was on friendly terms with Himmler's mother.

Many of Heisenberg's colleagues had long since left Germany by this time; after the unanimous adoption of the Nuremberg Race Laws by the Reichstag in 1935, there were no longer any Jewish physics professors in Germany.

Although the German government did not unconditionally embrace the ideas of Lenard and Stark, nor did it go so far as to prevent German scientists and engineers from exploiting quantum mechanics and relativity technologically, but the anti-Semitic agenda did destroy the Jewish scientific community in Germany. This attitude toward "Jewish physics" would work against Germany in the Second World War, because their now purely "Aryan physics" was only crowned with moderate success. The Nazis lost the race to develop rocket propulsion, radar, deciphering machines, and finally the atomic bomb. In the "physicists' war", Germany proved to be technologically inferior to its opponents.

Science in the Soviet Union

The situation in Russia was similarly ideological. After the Russian Revolution of 1917, science and technology were accorded enormous importance in the Soviet Union. It was the first state that wanted to base its future solely on scientific principles; the belief in progress was vehemently propagated. Science was supposed to modernise the country, which had been destroyed by world and civil war, and make it a great power again. At first glance, the Soviet Union seemed to be a paradise for scientists. But here, too, ideology was more important than scientific methodology.

Stalin personally supported the Soviet biologist Trofim Denisovich Lyssenko, who fundamentally rejected the idea of genes. For according to Marxist philosophy, the influences of the environment alone determine the development of man and society; just as communism educates people, plants and animals should also be malleable. The existence of genes, which set narrow limits to the variability of life, did not fit into this ideological scheme. Evolutionary theory and classical genetics were seen as an aberration of the capitalist West. In experiments that lacked any scientific basis, Lyssenko "proved" that plants are shaped by environmental conditions and that they can adapt to climatic conditions directly and without tediously long detours via mutation and breeding selection. The idea that plants can be "educated" within their lifetime cost two million people their lives. For from about 1930 onwards, Lyssenko had wheat varieties sown over wide areas of Siberia that were particularly productive in other regions but completely unsuitable for Siberian climatic conditions. According to the ideology of adaptability, they should have quickly got used to their new location. But things went wrong. The result was a further aggravation of the already frequent crop failures and famines. But even catastrophic failures did not stop Stalin and Lyssenko from continuing to play the direct adaptability card. The theory was applied with a similar lack of success to livestock farming, and until 1946/47 food supplies to the suffering population were effectively sabotaged in this way.

The system not only promoted those scientists who remained ideologically in line; it also destroyed those who professed scientific truth. Lyssenko had been supported for some time by Nikolai Vavilov, Russia's greatest botanist and geneticist, and one of the world's most important biologists of his time. In 1940, once again in need of a scapegoat for Lyssenko's failures, he was arrested and sentenced to death a few months later for sabotage and espionage. Vavilov died of malnutrition in January 1943 in a prison in St. Petersburg—renamed Leningrad at the time—the city where he had built up the world's first gene database from 1921.

Other great scientists also failed in the balancing act between science and ideology. Many of the Soviets' greatest technological achievements were made in special laboratory prisons, the Sharashkis, by imprisoned scientists. Stalin even got his atomic bomb in 1949.

* * *

The years between 1914 and 1920 were years of decline and collapse. Pessimism finally replaced the previous mood of optimism and trust in progress. The societies of Europe and America were traumatised by the war and on the verge of collapse. Scientific society, too, lay in ruins. The holes in the theories seemed impossible to patch up, and the previous fellowship among scientists had taken a serious blow. German and Austrian scientists were excluded from international conferences in the first years after the war as a result of the Treaty of Versailles, which placed the sole blame for the First World War on Germany, and German research contributions were no longer published at the international level. In some countries, ideological loyalty was valued more highly than scientific methodology. A scientific *community* committed solely to finding the truth no longer existed after the war.

The old science had played out. All seemed lost. And yet it was in precisely these disastrous years that the decisive beginnings of modern science were developed.

Part II

Geniuses Create a New World

Geniuses are characterised by more than just a particularly high level of talent or intelligence that elevates them above their peers: as creators of something completely new, they are creative in the true meaning of the word. They enter realms that are not only beyond what has already been explored and accomplished, but which their contemporaries did not even know existed. In this way, they expand the boundaries of the known and enlarge the playing field available to humanity.

Throughout world history, geniuses have almost always been solitary individuals. Kepler, Galileo, Newton, and Leibniz were beacons of their time, as well as James Clerk Maxwell, Ludwig Boltzmann, and Sofja Kowalewskaja in the second half of the 19th century. But in the first half of the 20th century, a strikingly large number of outstanding minds began working together across all the scientific disciplines, engaging in a lively exchange with each other. In physics and mathematics, chemistry, and biology, but also in philosophy and psychology, they opened up possibilities for their disciplines to deal with the fundamental contradictions that had arisen, some of which in antique times. In some cases, they even succeeded in eliminating the discrepancies. Their courage and achievement cannot be overestimated.

The first steps were taken by physicists such as Einstein, Bohr, and Heisenberg, when they left classical physics behind with relativity and quantum theory and thus moved outside the confines of Western philosophy, which had never been questioned for 2500 years and had been shaped by Plato and Aristotle. Mathematical geniuses such as Cantor, von Neumann, and Gödel also left the safe and familiar shores of classical mathematics and ventured out into deep waters. The first act of genius was to mathematically tame the infinities. They left behind the clarity and classical logic that had given their subject a foothold and framework until then, and conquered areas of mathematics whose apparent absurdity and high degree of abstraction defy any everyday experience and conceptuality. They even succeeded in taming chance.

In addition, the science of life received decisive stimuli from the early 20th century onwards. It was not least physicists and mathematicians who solved the riddles left behind by Darwin and Mendel in biology and advanced the decoding of heredity and the genetic code. The decisive breakthrough came in 1953 with the deciphering of the DNA code by the physicist Francis Crick and the biologist James Watson (as well as the much-neglected Rosalind Franklin). The resulting genetic engineering possibilities continue to occupy us to this day. With the discovery of neurons and synapses in our brain, an understanding of the human mind began to emerge, and also the possibility of influencing it.

From the late 1930s and early 1940s, later driven by the Second World War, the scientific centre of gravity underwent a geographical shift from Europe, with Germany at its centre, to the USA. This brought another group of geniuses onto the scene, such as Richard Feynman, who developed the first coherent quantum field theory, Robert Oppenheimer, (scientific) head of the Manhattan Project to build the first atomic bomb, or John Bardeen, who initiated the miniaturisation of computers with his semiconductor theory— and later also explained superconductivity, for which he received a second Nobel Prize in Physics. With this shift of the scientific centre, something fundamental happened in science: the American researchers ensured a fundamental change in the nature of all scientific disciplines, which still determines the nature of science today—starting with physics: Physicists in the USA left the field of philosophical reflection, which had played such a prominent role in European science from its origin until then. Instead, they now devoted themselves much more to the possibilities for concrete application of the new scientific findings. The triumphant advance of the technologies that came out of the scientific discoveries defining our world to this day had been set in motion in the US: electronics, digital technologies, lasers, mobile phones, satellites, televisions, radio, nuclear technology, medical diagnostics, new materials, genetic engineering processes, and neurotechnologies. Last but not least, a whole new mathematics, whose computational methods are so powerful that even global climate models have become possible, together with the miniaturisation of processors made possible by quantum physics, has led to the development of the computer.

It is still a crazy contradiction today: although we do not understand what exactly and why something happens in the subatomic world, we can calculate it exactly, and accordingly, we can also control it technologically. Plus, some core philosophical questions remain unanswered by science, while still being influenced by it, questions such as the limits of reason, the implication of randomness, the existence of reality outside the observer, and last but not least the nature of the human Ego.

6

The New Boys' Physics—A New Generation Discovers the Abstract World of Quanta

A chasm had opened up between the physics of the macrocosm and the completely contradictory behaviour of atoms and light particles in the microcosm. On the one hand, there was classical physics with Newton's mechanics and Maxwell's electrodynamics, on the other hand, Ernest Rutherford's atomic model, Planck's quanta, and the particle properties of photons according to Einstein. It was not yet clear which scientists were "on the right track": those who trusted the proven models and theories of physics, or those who suspected that some phenomena would probably not be explained in the future with the laws that had been known and proven for centuries?

It was in this situation that the young Danish physicist **Niels Bohr** came upon the scene. The brilliant theorist had gone to Manchester in 1912, just one year after completing his studies, to work with Ernest Rutherford on his new atomic theory. In his mid-twenties, he had the same age as Einstein when he had made his greatest discoveries seven years earlier. The problem with Rutherford's model was obvious: Why don't the negatively charged electrons immediately crash into the positively charged nucleus? Bohr had the decisive insight: if for some reason, according to Planck and Einstein, the energy of electromagnetic radiation is only emitted in quantised form, couldn't this quantum idea also be applied to the energy of the electrons in Rutherford's

This chapter is partly based on the book published by Springer-Verlag in 2018: Lars Jaeger, *The Second Quantum Revolution*, Springer (2018).

atomic model? Bohr introduced ad hoc hypotheses that once again tried the patience of the followers of classical physics:

- In their movement around the atom, electrons can only orbit on fixed paths. Contrary to Maxwell's laws, they do not radiate any energy without external influence.
- Analogous to Einstein's model of photons, the energy radiation and absorption of electrons can only take place through jumps between two neighbouring orbits, so-called quantum jumps.
- The transition from a state of higher energy to a state of lower energy (or vice versa) occurs with the emission (or absorption) of a Planck quantum or an Einstein photon.
- There is an orbit that represents the state of an electron with the smallest possible distance from the electron to the nucleus. The electron cannot get closer to the nucleus than on this path. Therefore, it cannot fall into the nucleus.

Ad hoc hypotheses describe and explain phenomena, but have no theoretical basis. There are three outcomes for them:
- They turn out to be wrong at some point.
- The theory will be provided later.
- The decision about true or false remains in the balance for a long time.

Now, one can object to Bohr's theory that it required no great art to think of the world with such a lot of imagination that it must necessarily fit the observations. His construct was so outlandish that it could hardly be taken seriously. But it was precisely this new quantum hypothesis that provided a solution to a puzzle that physicists had been unable to explain for over a century: that of the emission spectra of chemical substances. Independently of each other, William Wollaston in 1802 and Joseph von Fraunhofer in 1814 had observed that the individual chemical elements emit light at certain specific frequencies when heated, thus showing a barcode-like colour pattern. For example, the main lines of the element potassium lie at the frequencies 786 nm (red) and 404 nm (violet). Conversely, since it was possible to draw conclusions about the composition of materials based on their characteristic spectral lines, spectroscopy became an important tool in chemical analyses. No one knew why this was possible. Bohr's atomic model finally explained the mysterious spectral lines: in each atom, the electrons can only emit light at the frequencies that correspond to the energy difference between two allowed quantum states in the atom. Since atoms of different elements allow different energy states for their electrons, each element has its own characteristic emission spectrum. The pattern in the spectrum of each element is therefore a direct expression of the quantum nature of the electron energies in the atom!

After Planck's quantum hypothesis and Einstein's photon hypothesis, there thus appeared a third hypothesis from 1913 with Bohr's atomic model, which assumed a strangely erratic quantum nature in the microcosm. But the theoretical basis was still missing. The ideas about the quantum nature of the microcosm were floating freely in space, so to speak, without any connection to known laws of physics. Bohr and Planck themselves did not really believe in their quantum hypotheses, and only Einstein suspected that they were more than just an embarrassment.

For ten years there was no progress, this path seemed to have been a dead end. But then things started moving again in the early 1920s. Brilliant young theoretical physicists no longer tried to fit atoms into Newton's world picture, but developed entirely new, sometimes very abstract ideas for the atomic realm. Most of these physicists were born after 1900, so they were barely older than 20 when, far more carefree than Planck, Boltzmann, or even Einstein, they used their abilities, their spirit of optimism, and their considerable motivation to explain the world in a completely new way. Two places developed into the centres of the new physics: Copenhagen, where Niels Bohr now taught, and Göttingen, where the mathematician and physicist Max Born—a good friend of Einstein—was a professor. The Dane and his German colleague, three years his senior, became the intellectual fathers of a new generation of scientists. Many established physicists, still attached to the nineteenth century way of thinking, spoke disparagingly of the "new boys' physics".

Ups and Downs

Born in 1900, **Wolfgang Pauli** was one of the young men who rallied around Niels Bohr in Copenhagen (he had previously been an assistant to Max Born in Göttingen). During his studies, he had already dealt with the question of why there are many different orbits in Bohr's atomic model and not just one. With every "jump down" electrons give off energy, so after a certain time one should expect all electrons to accumulate on the lowest orbit—if for unknown reasons they do not crash into the nucleus. To resolve this inconsistency, Pauli introduced two new ad hoc hypotheses into the quantum world in 1925:

- An orbit occupied by an electron in an atom cannot be occupied simultaneously by other electrons for which the three attributes of energy, orbital angular momentum, and magnetic moment have the same properties. He later received the Nobel Prize in Physics for this idea, known as the Pauli exclusion principle.
- Because it was known that *two electrons* in the same quantum state could be found in one orbit, something that could be measured, Pauli postulated a fourth state variable called "spin". This intrinsic angular momentum had to have two possible states, referred to as *up* and *down*. With the rule that two electrons with different spins can travel on the same path even if they have the same energy, the same orbital angular momentum, and the same magnetic moment, the gap between theory and experiment narrowed.

Like almost all ad hoc hypotheses, Pauli's ideas raised more questions than they had answered. The spin hypothesis agreed with experimental observations, but what was the spin of a particle anyway? In the meantime, no one was really sure whether an electron was a particle at all (only then would an intrinsic angular momentum be conceivable). This was because, a year earlier, the young French physicist **Louis de Broglie** had put forward another ad hoc hypothesis in his doctoral thesis: if, according to Einstein, light can be understood not only as a wave but also as a particle, it should be possible to apply this dual nature to the electron as well! So, what happens if you assume that an electron is a particle but also a wave at the same time? This idea brought new momentum into experimental laboratories, because if you know what you are looking for, it is easier to find it. Within a very short time, it was experimentally proven that electron beams can indeed exhibit wave properties such as diffraction and interference. Incidentally, one of these experimental physicists was the son of the discoverer of the electron. The fact that his father Joseph John Thomson had received the Nobel Prize in 1906 for discovering the electron *as a particle,* and that his son George Thomson was awarded the same prize for demonstrating its *wave properties* in 1927 shows how physics was moving forward in leaps and bounds at that time.

The four properties ("quantum numbers") of an electron in quantum theory:
- The **energy** of the electron is the "principal quantum number" of the atom.
- The **magnetic moment** reflects the properties of the magnetic field generated by the electron moving in a circle around the atomic nucleus. It is quantised, i.e., it can only assume certain values.
- The **orbital angular momentum**, which describes the state of motion of a rotating body, is also quantised.

- Since an electron has no spatial extension, the quantum of its **intrinsic angular momentum (spin)** has no equivalent in the world as we are able to imagine it.

With Hay Fever to Helgoland

Things were getting crazier and crazier: light waves can appear as matter and matter can have wave properties. The more one learned about the quantum world, the more everything seemed to blur together. Physicists began to suspect that Newtonian mechanics could not be applied to the atomic world, but their *idea* of atoms was still based on their experience in the world of human dimensions. A simultaneous wave and particle nature overwhelmed people's imagination. They had hitherto transferred the familiar and descriptive world of imagination to the atomic world: like cannonballs, electrons should also have clearly measurable properties such as a position, a velocity, and a momentum. How can a massive sphere, no matter how small, have wave properties?

Unlike older physicists—now including Einstein and Planck—the young generation of physicists was prepared to accept that the conceptual and visual world of classical physics could definitely not explain the new experimental phenomena. The first step towards a completely new physics was taken by another theoretician from the Copenhagen group, called **Werner Heisenberg**, born in 1901. He was the first to dare to leave the variables of position and velocity of the electrons completely out of the description of atomic processes. No one had ever succeeded in measuring these parameters exactly anyway. Heisenberg only took into account the variables that could be determined in experiments, such as frequency, energy, and intensity, and tried to describe the properties of the atom unambiguously with these alone.

With this approach, Heisenberg was the first person who no longer tried to form a *picture* of the events occurring in an atom. He relied solely on mathematics. He found that he could not achieve his goal with the classical computational tools such as addition and multiplication. He thus developed a complicated calculational method in which abstract operators were used to calculate, so that definite numbers and variables no longer came out as solutions, but were replaced by ranges of numbers and variables. According to human intuition, this kind of mathematics in which one worked with axes instead of scalpels, so to speak, should lead to rather indeterminate results. But it turned out that Heisenberg's new calculational method was able to reproduce and predict with great accuracy all the properties of atomic systems that could be measured in experiments.

The night Heisenberg made the decisive breakthrough, he was on Helgoland, where he had travelled in June 1925 because of his nagging hay fever. In his euphoria, Heisenberg, an enthusiastic and excellent climber, ran out of the house and scaled the Lange Anna, Helgoland's landmark. Fortunately, he survived this daring climb, saving not only his own skin, but also his brand new insight into the quantum world. Back in Göttingen, the 23-year-old presented his work to his fellow researchers. His colleague Pascual Jordan, who was a year younger, and his professor Max Born recognised Heisenberg's calculational method to be based upon a so-called matrix algebra. This method, which every mathematics student learns in their first semester today, was almost unknown at the time, even among experts. So, Heisenberg had invented the wheel for the second time. Heisenberg, Born, and Jordan perfected the "matrix mechanics" as a trio and used it for the first time to bring a certain order to the confused jumble of ad hoc hypotheses then predominant in quantum theory.

However, this step came at a price: with Heisenberg's matrix mechanics, it became clear why physicists had not previously been able to precisely determine the position and velocity of an electron at the same time. It had not been due to inadequate experiments, but the nature of the atoms themselves! The more precisely the position of an electron is determined, the less is known in principle about its velocity and vice versa. This is precisely what Heisenberg's most famous formula, the "uncertainty principle", says:

$$\Delta x \cdot \Delta p \geq h/4\pi.$$

Here, Δp and Δx denote the inaccuracy in the measurement of the momentum and the position, respectively, and h stands for Planck's quantum of action. Heisenberg had thus shown mathematically that the properties of "position" and "velocity" of electrons are connected to the measurable properties (see above), but cannot be determined exactly in the atom.

Chance Shows up Again

Another urgent task of the new physics was to provide a mathematical description of the smallest building blocks of nature—light, electrons, electromagnetic waves, etc.—both as particles and as waves in *a single* formula. For particle and wave properties still had to be considered separately.

- (Electromagnetic) waves can be described with Maxwell's equations.
- "Classical" particles are described by Newton's mechanics and Ludwig Boltzmann's statistical mechanics (extended by Einstein).

For several years, attempts had been made unsuccessfully to describe light mathematically, taking into account both its wave and particle nature. De Broglie's idea opened the way to another solution: instead of starting from Maxwell's equations and somehow adding the properties of matter to the wave nature of light, the task was now to find an extension of Newton's equations, which describe the behaviour of particles, to include their wave nature. So how could one represent mathematically the idea that a particle behaves like a wave?

As is so often the case in science, a precursor from long ago was found for precisely this task. As early as the first third of the nineteenth century, the Irishman William Hamilton and the German Carl Gustav Jacobi had described classical mechanics as a mathematical limiting case of wave mechanics. They could not have foreseen that their work would serve as an important foundation stone for a completely new physics a hundred years later.

The **Hamilton–Jacobi equation** represents the motion of a particle as a wave. It had been a long-cherished goal of theoretical physics to find an analogy within classical mechanics between the propagation of light and the motion of a particle. This desire went back to the eighteenth century; among others, the mathematician Johann Bernoulli had been interested in this problem.

The wave equation of Hamilton and Jacobi is similar but not identical to Schrödinger's equation, which is why the Hamilton–Jacobi equation is considered today to be only a first, accidental approximation of classical mechanics to quantum mechanics.

It was not one of the "young savages", but the somewhat older Austrian **Erwin Schrödinger**, born in 1887, who—starting from the Hamilton–Jacobi equation—found the long-sought equation for a wave theory of the electron. For the mathematically savvy reader, Schrödinger's equation was mathematically more complicated than Maxwell's classical wave equation, since it only involved the first derivative with respect to time.

The following anecdote shows that physicists and mathematicians are also people of flesh and blood: Schrödinger made his breakthrough when he actually wanted to relax in unknown female company during a skiing holiday; his wife had stayed in Zurich. When he returned home and worked out his formula in more detail, he was supported by his colleague Hermann Weyl, who was not only his close friend but also his wife's lover, something he tolerated.

Schrödinger's ingenious idea was to describe the electrons as standing waves whose frequency determines the path length and thus the distance from the atomic nucleus. One piece of the puzzle after another fitted into the picture.

- What Bohr's model still described as quantum leaps, Schrödinger's wave mechanics now turned into transitions from one natural electronic oscillation state to another.
- Because electrons bound in atoms can only assume certain wavelengths or frequencies, their energy can only have certain values, otherwise their waves do not "fit" the length—figuratively speaking—of their circular path.

With his wave equation, Schrödinger had found another theoretical basis for describing what happens in the microcosm. Until now, physicists who were on the trail of electrons had done little more than stumble around in the fog with their ad hoc hypotheses. With Schrödinger's equation, the behaviour of these quantum objects in the atom could be explained precisely, elegantly, and entirely in accordance with Bohr's atomic model. It was a decisive building block on the way to a consistent theoretical foundation for the new quantum theory. For the first time, one could cherish the hope of being on the right path.

Dispute Among Physicists

In theoretical physics, therefore, there were now two completely different explanations for what was happening in the atomic world:

- Heisenberg's matrix mechanics relied on pure mathematics, without taking into account whether the result could be visualised or not. Einstein called this matrix mechanics "witchcraft".
- Schrödinger's wave mechanics was less abstract and matched Bohr's atomic model.

Which should one choose? The physicists disagreed and camps formed. Heisenberg and Schrödinger took the matter personally, and serious competition developed between them. Above all, the twenty-year-old Heisenberg was irreconcilable towards this competitor, 14 years his senior. While Schrödinger believed there was a common ground between their two theories, Heisenberg openly and even tearfully fought Schrödinger's approach, hoping that errors would be found in his equation. He wrote to his friend Wolfgang Pauli about Schrödinger's wave mechanics:

The more I think about it, the more heinous it seems.

It may have contributed to Heisenberg's bitterness that most physicists preferred Schrödinger's less abstract interpretation. Yet the theory put forward by Born and Schrödinger and Heisenberg's idea were not that far apart.

- Strictly speaking, Schrödinger had also abandoned the idea of an electron orbit and replaced it with natural oscillations of waves.
- As in Heisenberg's interpretation, there is no longer an exact position of the electron in Schrödinger's interpretation.

The uncertainty principle of Heisenberg's matrix mechanics and Schrödinger's interpretation of the wave character of electrons proved to be perfectly compatible on closer inspection. Only a few months after Schrödinger published his equation, Max Born formulated a new interpretation of Schrödinger's equation, according to which Schrödinger's waves do not describe the physical motions of electrons per se, but the distribution of the *probability* of an electron being at a given position at a given moment. When many electrons are considered, the mathematical probability distribution of individual electrons becomes the distribution observed in experiments, wherein a large ensemble of electrons can exhibit wave properties—meanwhile the individual electron remains a particle in mathematical terms. Although Born had built a bridge between Heisenberg and Schrödinger with this interpretation, Heisenberg, who was now working for Bohr in Copenhagen, accused his former superior of having "defected to the opposing camp".

Soon afterwards, Schrödinger was even able to prove that his own wave mechanics could be derived mathematically from Heisenberg's matrix mechanics and, conversely, Heisenberg's from Schrödinger's equation. So, what had at first looked like mathematical descriptions of two completely different phenomena turned out to be *mathematically equivalent!* This agreement made physicists increasingly confident in the new theory, which they now generally referred to as "quantum mechanics".

A Farewell to Classical Certainty

Astonishing parallels had also emerged between electrons and light quanta:

- Einstein had shown that a large number of light quanta take on the properties of an electromagnetic wave, while the individual photon is explained as a particle.
- According to Born and Schrödinger, individual electrons were also particles that assumed wave properties in large ensembles.

But while Einstein had used the classical methods of Maxwell's equations and statistical mechanics and thus remained within the boundaries of classical physics, Schrödinger and Heisenberg broke new ground with their equations, and Born likewise with their interpretation. This is because the fuzziness in the position and momentum does not allow any statement about where exactly an individual electron is travelling on its path, nor at what speed or with what momentum. Although the wave function of a quantum system obeys deterministic dynamics (Schrödinger's equation is a deterministic differential equation), the variables can only be determined statistically—quantum mechanical states can therefore only be assigned probabilities. Chance thus became a decisive component of the new physics. Einstein could not accept the idea of this indeterminacy and, as the following chapter shows, spent his life searching for a way to resolve the ambiguity of the quantum world and thus return it to the safety of classical physics.

Despite this break with classical physics, it was now possible to describe atoms in a mathematically consistent way within the framework of the new quantum theory; the results of the calculations matched experiment results exactly. However, the new quantum mechanics also demanded quite a bit of conceptual adjustment:

With the new mathematical tools, there was no longer any reference to anything concrete or explicit.

The electron was no longer described in a clearly imaginable three-dimensional space, but in the abstract state space of a wave function, which can be interpreted mathematically as a space with an infinite number of dimensions. Mathematicians had already developed a suitable framework for this with the so-called Hilbert spaces.

> The three-dimensional space in which we humans find our way around has three basis vectors: height, width, and depth. **Hilbert space,** on the other hand, is an infinite-dimensional space with an infinite number of basis vectors. As a vector in such a space, each wave function can be expressed as a linear combination of infinitely many basis vectors. So while a point in classical space needs three coordinates to be fixed, in the case of wave functions, there are infinitely many.

The values of the wave function were not limited to the familiar real numbers. Because it is a function with complex number values, it also includes a part that would be impossible as a physically measured value. Only its squared modulus, which is always real, can be interpreted physically, indeed, as the probability distribution of the given state.

Schrödinger himself was most dissatisfied with this development in the end. Other physicists, on the other hand, especially those in Niels Bohr's group in Copenhagen and also Werner Heisenberg, who had meanwhile gone to Leipzig but was still closely associated with Bohr, enthusiastically embraced the new findings and tried to develop quantum mechanics further. One of the pieces of the puzzle that did not yet fit into the overall context was wave–particle duality. Neither Heisenberg's matrix mechanics nor Schrödinger's wave theory could explain *why* this phenomenon existed. On the contrary, a major conflict about the nature of the quantum world was just getting under way. However, an experiment would soon shed light on the matter and in doing so become famous around the world.

> **Complex numbers z** have the form $z = x + iy$, where x and y represent real numbers and i is the square root of -1, impossible in the world of real numbers.

The Confounded Double-Slit Experiment

In our everyday lives, we often have to settle for probabilities. For example, if there is a table in one of two locked rooms and we do not know in which of the rooms it is located, we can guess with a probability of 50%. Probability comes into play because *subjectively* we do not know which room houses the table. In the microcosm, however, probabilities are an *objective* component of quantum mechanical dynamics. If the table behaved like a quantum particle, it would be in both rooms at the same time and in neither of them at the same time. Only at the moment we look does the quantum table "materialise" in one of the two rooms.

This is the phenomenon of superposition of different possible states of a quantum particle. Such superpositions exist not only with regard to the position of a quantum object, but also for other properties:

- whether it behaves like a wave or a particle,
- what spin it has,
- what position it occupies and what momentum it has.

The effects of superposition are particularly impressive in the famous double-slit experiment. Here, an electron or photon beam hits an aperture in which two narrow, parallel slits are cut. A certain proportion of the particles pass through the slits and collide with a photographic plate mounted behind the aperture. Each impacting electron (or photon) leaves a black dot on this plate. At first, they seem to behave like macroscopic particles: they fly through either the left or the right slit and accordingly black dots accumulate behind the apertures to form two stripes. After a sufficient number of such hits, however, it becomes apparent that what is happening is somewhat more complicated: over time, an interference pattern of alternating black and white stripes becomes visible on the photographic plate. Such patterns are known from wave optics and can also be observed in light or water waves passing through a similar setup.

> **Superposition**: The exact state of a quantum particle is not only *subjectively* unknown, but also *objectively* undetermined. Only when measured does it fall into a specific state with clearly defined, distinguishable properties. However, there are exceptions: the Heisenberg uncertainty principle states that, for example, the speed and position of an electron can never be known at the same time.

Physicists had an explanation for this effect: electrons behave like *waves* on their way through the slit and behind it. Only when an electron hits the photographic plate does its particle nature come to light. Before that, its exact location is basically unknown. Because a wave is not localised but distributed in space, it is also impossible to predict where it will hit. At places where the *wave amplitudes* of the electrons have increased when passing through the two gaps, the electrons are more likely to hit as *particles* than where their waves almost cancel each other out. This is precisely the behaviour predicted by Born's (or for photons, Einstein's) interpretation of probability waves.

But when the physicists carried out the same experiment with *single* electrons fired at intervals one after the other, they would have a big surprise. Although it was now impossible for the successive electrons to interact with each other and thus affect each other's wave nature, the interference effect still occurred. There were only two possible explanations for this:

1. The individual electron particles somehow "know" how they have to behave in order to create a wave pattern.
2. Each individual electron passes through both slits at the same time as a wave and then interferes with itself.

Because the first interpretation seemed even crazier than the second, physicists preferred the second explanation. They now tried to detect individual electrons that had passed through one aperture or the other using a laser beam. Would they be able to measure "half" electrons? Again, the experiment revealed something astonishing:

- If attempts were made to measure, between the slit and the photographic plate, the slit an individual electron passed through, a clear result was obtained: it came through either the left *or* the right slit, not through both slits at the same time. Apparently, the first explanation had been the right one after all.
- If a sufficiently large number of electrons were fired one after the other and their trajectory measured behind the slit, two clearly delineated lines appeared on the photographic plate, i.e., there was no longer an interference pattern. (If *no* measurement was made with the same apparatus, the familiar interference pattern reappeared.) It was as if the observation of the electron destroyed its wave nature and revealed its particle nature. These experimental results from the quantum world contradicted all experiences people have in the macro world. The measurement that took place after the passage through the slit *retroactively* influenced the particle! But how did the electron "know" that it would be measured *later* and therefore already had to behave at the slit, not like a wave, but like a particle? And how did it "know" that it would *not* be measured later and that it should therefore behave like a wave at the slit?

Experimental physicists came up with an experimental setup that took the whole thing to the extreme:

- Individual electrons are measured after their passage through the two slits.
- Through the measurement, they lose their wave character. Without further intervention, two separate black lines appear on the photographic plate, coinciding with the aperture slits.
- With a so-called quantum eraser, the information gained during the measurement is destroyed again before it is passed on to the observer—and before the electron hits the photographic plate.
- The effect is that the electron behaves as a wave again; despite the measurement, an interference pattern is created. The *subsequent* annihilation of the information reverses the change in the quantum object caused by the measurement.

Physicists now had to come to terms with all this. It was not the measurement itself, but only their knowledge of its result that determined the properties of the quantum particle, and that too retroactively in time. It was clear that there were still huge gaps in our attempts to explain the nature of the electron!

> **Quantum eraser**: When photons are measured, suitable crystals determine whether they have flown through the left or right slit. A downstream polarisation filter, which only allows certain oscillation directions of the light to pass, immediately deletes this information.

Dirac's Stroke of Genius

Superposition had even more amazing effects in store for us. For example, consider the exact moment when the electron in the double-slit experiment hits the photographic plate:

- Shortly before the impact—the time period can be chosen as small as desired—the electron wave still exists as a superposition of different spatial states. The electron can therefore strike at many different places on the photographic plate.

- When it hits, it ceases to exist as a wave without any time delay, it is *a* particle that leaves behind *a* black dot. The previously objectively indeterminate place of impact has become a definite place. It is now impossible for it to hit somewhere else.

Physicists speak of an "instantaneous collapse of the wave function," where instantaneous means "without any time delay". How can this instantaneous change from wave to particle properties be explained?

The Schrödinger equation could describe and interpret the previous experimental results very well, with the exception of the still ad hoc formulation of the existence of spin. But it could not depict the fact that an electron, which as a wave still had innumerable possibilities for hitting somewhere on the photographic plate, could change into a state, without any time delay, in which it would leave a black spot at a certain point, whereupon all other possibilities seemed to be instantaneously erased. This effect was not compatible with Einstein's theory of relativity.

In 1927, in a single stroke of genius, the English scientist Paul Dirac, only 25 years old, succeeded in finding the cause of this from purely theoretical considerations, and hence in solving the problem. Schrödinger's equation still assumed time to be independent of space. Thus, he wrote in 1928:

> The incompleteness of the previous theories lie in their disagreement with relativity, or alternatively, with the general transformation theory of quantum mechanics.[1]

Dirac transformed the equation so that it took into account the spacetime of relativity. Now the instantaneous events of wave collapse became mathematically representable, and there were two further gifts for physicists:

- Without any additional assumption being necessary, the existence of spin arose from it all by itself. The spin hypothesis, which Pauli had been compelled to introduce, turned out to be a direct consequence of extending the quantum theory by the consideration of space–time.
- When a more detailed quantum theory of the photon became known twelve years later (quantum electrodynamics), it was finally also possible to derive the deep connection between spin and quantum statistics that had led Pauli to his exclusion principle.

---

[1] From Dirac's paper *The Quantum Theory of the Electron*, Proceedings of the Royal Society (1 February 1928).

Scientists dream of such moments of happiness, when important pieces of the mosaic suddenly fit together to make a coherent picture.

While the theoretically coherent derivation of electron spin made physicists cheer, they were shaken by another consequence of Dirac's equation: it allows for the existence of electrons with negative energies!

- In the Schrödinger equation, which does not take spin into account, there is only one wave function.
- With Pauli's consideration of the spin, this became a wave function with two components (called up and down).
- The even more complicated Dirac equation has four spatial wave functions as solutions, known as spinors. Two of the components describe particles in two different spin states, each with positive energy, the other two describe two spin states with negative energies.

In classical physics, negative energy is an impossible, even unthinkable thing. The idea of a car with negative energy illustrates this: it would not move backwards, for example, but would move less than not at all. How should physicists interpret the two negative-energy solutions of the Dirac equation? Could they simply ignore them?

The problem of negative energies actually already occurs with Einstein. His equation $E = mc^2$ only applies to particles that are at rest. If they move, this equation is extended by further terms in which the energy appears as a square:

$$E^2 = p^2c^2 + m^2c^4$$

The equation for E thus always has two solutions, one with a positive value, one with a negative value. The solution with a negative value was ignored because a negative energy was unimaginable.

A negative energy reverses the signs of other physical quantities: an electron with negative energy would have the same mass and spin, but would carry an opposite magnetic moment and a positive charge. Dirac interpreted particles with negative energy as "anti-electrons" three years later, in 1930. What physicists did not know at the time was that the Soviet physicist Dimitri Skobeltsyn had already discovered precisely this particle just a year earlier. In an attempt to measure gamma radiation as a component of cosmic radiation in a cloud chamber, he had observed particles that behaved like electrons, but whose paths were curved in exactly the opposite direction in an applied magnetic field. Thus, with identical mass, they possessed a charge opposite to that of the electron. Unfortunately for Skobeltsyn, he did not pursue this phenomenon any further—he would have won the Nobel Prize. Frédéric and Irène Joliot-Curie (Marie Curie's daughter) also came across evidence of "positive electrons" early on, but interpreted them as protons.

It was the American physicist Carl David Anderson who, on 2 August 1932, demonstrated the existence of the anti-electron with an apparatus very similar to the one Skobeltsyn had used three years earlier. He called it the positron. For Dirac, the detection of the positron was his greatest triumph, and for theoretical physics it was a great moment. A new branch of physics was born: particle physics.

7

Einstein and Schrödinger Versus Bohr and Heisenberg—How Philosophy Was Displaced by Mathematics

For a long time, it was up to religion to explain the world: God (or gods) made the world and imposed his (their) laws. Scholars of the European Middle Ages wanted to understand these laws in order to get closer to the work and will of God. There was no room for other explanations. As late as 1600, the Dominican monk Giordano Bruno, who imagined the universe as infinitely large and without a creator, was burned as a heretic. But not long after, the time of Galileo and Kepler dawned. They, too, felt the pressure of the Church, but gradually religion lost its explanatory power in the Western world. Insights and reflections of ancient philosophers, which had long been "approved" only in part, began to determine thinking again. In addition, there was a new and important factor whose value had been little appreciated in antiquity: the systematic observation of nature through experiments. Thus, it came about that from the beginning of the seventeenth century people were able to look at the world in a completely new way. This was the birth of the classical natural sciences. Many of the physicists of the time were religious (and some of today's still are), but when it came to their subject, they now no longer adhered to mystical attempts at explanation, but to theories that were rationally comprehensible, internally consistent, and above all demonstrable by experiment.

For more than three hundred years, until the 1930s, classical physics and classical philosophy going back to the Hellenistic tradition—supplemented by modern philosophies, such as Kant's—were inextricably interwoven in

L. Jaeger, *The Stumbling Progress of 20th Century Science*, https://doi.org/10.1007/978-3-031-09618-1_7

the Western world. In this period, it was natural for physicists to participate in philosophical discussions. Conversely, many philosophers were able to discuss deep connections in the natural sciences; Kant, for example, also taught physics at his university. It was not until the middle of the nineteenth century that mathematical calculations became so complex that only a few philosophers were still able to comprehend the scientific advances of human knowledge. Physicists, on the other hand, maintained their interest in philosophy; Heisenberg, for example, was fascinated by Plato, and Einstein by Spinoza and Kant. But one particular facet of Western thought, shaped by Plato and Aristotle, was to cause the physicists a lot of trouble: since ancient times, people have assumed a dualism that tears the world in two.

Physicists are actually philosophers, because they want to get closer to the basic human question "How does the world work? They are not only concerned with questions of physics itself ("How does a pulley work?"), but also with metaphysics, i.e., with what goes beyond observable and measurable physics.

- Behind all the phenomena we perceive there is something absolute, so-called **substance**. It is independent of our observation, indestructible, and unchanging, and thus the original basis of all things. For a long time, certain basic entities were considered to be substances, including fire, water, earth, and air. Newton also searched for the basic substance of the world, in the form of natural laws that are valid everywhere and eternally.
- **Accidents**, on the other hand, are variable. Their properties are neither concrete nor clearly describable, and they are sometimes even random (which explains classical physicists' great aversion to chance). The category of accidents also includes the impermanent, subjective impressions we have of the world through our senses, including colours, sounds, and shapes.

Substance: that which is objective, unchanging, and absolute.

Accident: that which we subjectively perceive as changeable and random.

The idea that the world is divided into substances and accidents still determines our perception of the world today.

The fathers of classical physics from Galileo to Kepler to Newton had grown up in the philosophical tradition of clearly distinguishing between substance and accident. They did not even ask themselves whether it was permissible to transfer this separation from philosophy to their field. Dualism even became *the basis of* classical physics in general. For these were precisely its scientific goals:

- to find the material, unchanging building blocks of the world,
- to extract the substance of things in the form of eternally valid laws from the confusion of subjective impressions with the help of objective, repeatable experiments.

Through Newton's laws, people thought they were close to the *immaterial substance*, i.e., the all-encompassing laws of nature. Only a few details seemed to be missing to finally hold the all-explaining universal formula in their hands. When physicists proved the existence of atoms, they thought they were also within reach of the fundamental *material substance*. It was initially a disappointment that Rutherford's experiments showed atoms to be destructible and changeable. But then they got on the trail of even smaller particles: photons, electrons, and protons. Had they finally hit upon the sought-after substances that make up the universe? To answer this question, they tried to interpret what they knew about the world of quanta and to put it into a consistent theoretical framework.

A Vivid Explanation from Denmark

At the beginning of the 1920s, it was known that strange conditions prevailed in the quantum world. Niels Bohr and the physicists who had gathered around him in Copenhagen explained it like this:

- **Wave–particle duality**. Bohr spoke of the "complementarity of particle and wave", meaning that wave and particle properties of a quantum object are mutually exclusive and at the same time complementary—an electron is therefore a wave *or* a particle, never both at the same time.

- **Heisenberg's uncertainty**. Heisenberg had shown that the position and momentum of a quantum object cannot be exactly determined at the same time. The more precisely the position is determined, the more indeterminate the momentum is, and vice versa.
- In addition to the fuzziness of momentum and position, there are other **ambiguous properties of quantum objects**. If a physicist measures a quantum object *as a particle*, he must use an appropriate measurement setup and fall back on Heisenberg's matrix mechanics when calculating the particle properties. If, on the other hand, he wants to measure a quantum object *as a wave,* he needs a completely different experimental setup; mathematically, Schrödinger's equation with Born's statistical interpretation then comes into play. Because a quantum object cannot be measured as a wave and a particle *at the same time*, it cannot be a wave and a particle at the same time.

What is known today as the **Copenhagen interpretation** was not yet clearly defined in the 1920s. At a lecture in the USA in 1929, Heisenberg spoke of the "Copenhagen spirit of quantum theory". However, there is no text from that time that explicitly describes a Copenhagen interpretation. This name was probably only coined by Heisenberg in the 1950s, when alternative interpretations of quantum theory emerged and a clear designation became necessary.

This still quite vivid interpretation of the quantum world is called the "Copenhagen interpretation" of quantum mechanics in Bohr's honour. The bottom line was a picture of tiny particles without clear and unambiguous properties. This fact was explained—and to some extent still is—in such a way that, at the moment of measurement, the properties of the quantum object are changed. This implies that its original properties can never be measured. But the properties of quantum objects such as electrons, photons, etc., are not only indeterminate before measurement, *they do not exist at all*. To say it again quite clearly: a quantum object has no independent properties before the measurement! Only the measurement gives it any such properties.

The closer physicists came to the peculiarities of quantum objects, the further away the idea of a concrete, unchanging basic substance of the world became. Wherever they looked, everything solid, reliable, and familiar seemed to slip between their fingers.

The Copenhagen interpretation could boast great mathematical successes, but on a philosophical level it did not have much to offer. Its proponents, all members of a younger generation (except Bohr), had the attitude "What cannot be measured, we cannot know and therefore need not be considered." As long as mathematics led them to useful results, they were by and large satisfied. But other physicists, mostly older ones, were still strongly rooted in philosophy. They had to endure even worse than the bitter disappointment of not finding the substance of the universe in the quantum world: with quantum theory, other central concepts also disappeared from their mental map:

1. Reality
2. Causality
3. Identity.

An Unreliable World

Until about 1925, a basic assumption in philosophy was that there is **an objective reality that exists** independently of observation.

- In our everyday perception, things have concrete properties that are influenced neither by their surroundings nor by our perception—they are real. A tree is still there in the forest even if no one is looking at it. With the help of classical physics, the properties of an object, and thus also reality, can be clearly described and predicted.
- According to the Copenhagen interpretation, quantum particles do not possess any objective, independent properties, which is why they have no *reality*, but "only" a *potentiality*. The potential properties only become real when a concrete macroscopic object, such as a measuring device, interacts with the quantum object, whereupon the latter shows certain properties that depend on the nature of the measurement. Philosophers put it this way: quantum objects do not have their own substantial form of being.

For the followers of the Copenhagen interpretation, there was a reality only in the macroscopic world. In their research, they therefore no longer wanted to waste time looking for a material substance, independent properties, and real events in the quantum world. They relied solely on the abstract mathematical representation of quantum objects and thus threw two and a half millennia of Western philosophical history overboard.

If quantum particles have no properties, there is no **causality**. In classical physics, the properties of objects can be used to infer their behaviour. If you know the exact position, velocity, mass, air resistance, etc., of a cannonball, you can calculate where it is at any point in its flight and exactly where it will hit. Thanks to the measurable properties in our macro world, events run completely deterministically and are therefore predictable. In the quantum world, on the other hand, only certain aspects of the events can be calculated exactly. For example, we know what the half-life of a certain radioactive substance is, but the decay of any given atom within that substance happens without a trigger. The exact time of this event cannot be determined in principle.

Quantum Particles Without Identity

In addition to reality and causality, there is a third concept that physicists had actually assumed to be non-negotiable and yet nevertheless had to say goodbye to: **identity**. In our macro world, we define ourselves and all objects by our and their identity, that is, by the fact that we and they are unique and distinguishable from everything else. But even in this respect, a fundamental difference between the macro and quantum worlds becomes clear:

- Each blade of grass in a meadow has its own identity; you could give each individual blade a number. If you swap two of them, the new state of the meadow is no longer identical to the old one.
- In the quantum world, this is different: for example, the two electrons of a helium atom are not single, independent objects. Since they have no objective properties, they cannot be distinguished on the basis of properties. Even their spin does not make a difference, although Pauli introduced it in order to endow the two electrons in their atomic shell with different characteristics. In fact, all we know about the spins of two helium electrons is that both spin states are represented at any given time. When particle 1 is in the "spin up" state, particle 2 is in the "spin down" state, and vice versa (the concept of superposition is discussed below). Because the two electrons can be interchanged without changing the overall physical state, they have no identity of their own.

The indistinguishability of quantum particles contradicts the classical philosophical principle that Gottfried Wilhelm Leibniz had formulated 250 years earlier, but which Aristotle also knew: the *principium identitatis indiscernibilium* (abbreviated: *pii*). This "principle of the identity of indiscernibles" says that there cannot be two different things that are completely the same in every respect. Immanuel Kant pointed out that, for us humans, it is above all the factor of "place" that is significant: even if the properties of two parts punched out of the same material and by the same machine appear to be very much the same, they must at least be in different places.

The fact that *pii* does not apply in the microworld is not a theoretical gimmick, but the reason why fundamentally different statistics apply in quantum theory than in classical physics. The following example illustrates this.

Two balls collide with the same momentum from opposite directions and fly apart in different directions after their collision. If one of the balls happens to be deflected by exactly 90 degrees, the other ball will also be deflected by exactly 90 degrees according to the law of conservation of momentum, and it will then fly in the opposite direction.

- **In the macro world**: Assuming tennis balls are launched at each other many times, they will fly upwards at an angle of 90° just as often as they fly downwards. If, for example, one ball is deflected downwards by 90° 20 times in a certain number of collisions, then the other ball will fly upwards by 90° each time, i.e., also 20 times. In total, there is an exact 90-degree deflection 40 times after the two balls collide.
- **In the quantum world, we have the following situation**: In the case of quantum particles that are fired toward each other, the two possibilities (particle 1 flies upwards, particle 2 downwards, as well as the reverse possibility) coincide into one—the particles are, after all, indistinguishable. In the counting method of quantum physics, therefore, the probability that exactly one 90-degree deflection will be hit in a collision is only half as great as in the case of distinguishable classical particles. While tennis balls fly away 40 times with 90-degree deflection in the thought experiment, one would only be able to measure a total of twenty 90-degree bounces for quantum particles under analogous conditions. The human mind resists this finding, but in the quantum world, mathematics alone is called to account. This mathematical solution has been confirmed in a suitable experiment.

In the quantum world, it is even a little more complicated. The wave function of a system is called Ψ (Phi); according to Born, the result of a physical measurement is determined by the *square* Ψ^2 of the wave function. Therefore, if two components of a wave function are interchanged, two solutions result: Ψ can remain unchanged or the sign changes—for the value of Ψ^2, such a sign change is in any case without real consequence. And yet mathematics tells us that there must be two types of particle that behave differently when swapped. Some change their sign when swapped, others do not. This difference can actually be proven experimentally:

- In the case of **bosons**, the wave function Ψ remains unchanged when it is swapped. $+\Psi$ remains $+\Psi$ and $-\Psi$ remains $-\Psi$. These particles were named after the Indian physicist Satyendranath Bose who, together with Albert Einstein, formulated the first theory for this type of particles in 1924. Photons are also bosons.
- With **fermions**, the sign of the wave function changes when swapped. $+\Psi$ becomes $-\Psi$ and vice versa. This group of particles is named after the Italian physicist Enrico Fermi, who first described their statistical behaviour in 1926. Shortly afterwards, Paul Dirac created the theoretical basis for the fermions, which also include the electrons. Incidentally, the antisymmetry of the wave function of fermions is the reason for Pauli's exclusion principle, according to which two electrons can only assume non-identical quantum states, as we saw in the last chapter.
- The distinction between bosons and fermions makes the quantum mechanical counting method particularly complicated. The fact that a 90-degree event is measured exactly 20 times in the above collision experiment is a simplification. If the experiment is carried out with fermions, not a single 90-degree event is obtained; with bosons it is now 80. So, physicists not only had to come to terms with the fact that indistinguishable quantum objects have no identity of their own, they also had to learn to perform calculations that were completely different from the familiar ones.

The Fate of a Cat Becomes the Fate of Quantum Theory

The Copenhagen interpretation of Bohr and Heisenberg was highly controversial from the beginning. This was not least due to one of their core theses, according to which the macro world and the quantum world are strictly separated from each other. In his correspondence principle formulated in 1920, Bohr had stated pragmatically that the laws of classical physics apply to "sufficiently large" systems and the laws of the quantum world apply to "sufficiently small systems". Where exactly on the size scale the transition is located, he had to leave open. However, it was assumed that the transition does not happen gradually, but that there must be a sharp separation, called the Heisenberg cut. The alternative would have been that there is a region on the size scale in which two different sets of laws somehow apply simultaneously.

Schrödinger also recognised that the term "Heisenberg cut" only super-ficially concealed the central logical gap in the Copenhagen interpretation: somewhere on the size scale, the quantum world and the macroscopic world must touch, even though completely different laws must apply in each case. At this point, the laws of the quantum world merge into the laws of the macro world (and vice versa). Here, it must also be ensured that the laws of the quantum world, which are bizarre for us, do not "seep" into the macro world. How is that supposed to work when there are such conceptual differ-ences? And where exactly on the size scale is the Heisenberg cut supposed to be?

The **Heisenberg cut** designates the place on the scale of magnitude at which the quantum world, together with its laws, merges into the macro world of classical physics. Heisenberg's argument was that, without such a cut, quantum theory would ultimately have to be applied to the universe as a whole, in the form of a single wave function for the entire cosmos.

In 1935, Schrödinger described his famous thought experiment with the cat in his essay "The Present Situation in Quantum Mechanics". His aim was to reveal the absurdity of the Heisenberg cut. In a locked steel chamber there is a cat and a certain amount of a radioactive substance calculated in such a way that statistically one of its atoms decays every hour (since it is a kind of average, it may be that one, two, or even no atom decays in any given hour). Each decay of an atom is detected by a fine measuring device; if there is a decay event, it shatters a container of deadly hydrogen cyanide. After an hour, an observer will not know whether one of the atoms has decayed and whether the cat is alive or not. The genius of Schrödinger's thought experiment is that it directly causally couples the quantum and macro worlds and thus makes the strict separation postulated by Bohr and Heisenberg permeable: only one atomic nucleus has to decay for a macroscopic object—the cat—to be directly affected.

- As long as no measurement takes place on the atomic nucleus, it is in a state of superposition of "decayed" and "not decayed". Only through macroscopic measurement, i.e., by opening the steel chamber and looking into it, does the wave function of the atomic nucleus collapse, and only then does its objectively indeterminate state become a definite state.

- In Schrödinger's experimental setup, the state of the atomic nucleus is undetermined, and so therefore is the state of the cat in the steel chamber. This is now also supposed to consist of a superposition of "dead" and "alive": As long as the door to the box is not opened, the cat is *objectively* in both states at the same time! This sounds absurd to our everyday imagination. How can it be dead and alive at the same time and only assume one of the two states when we look at it?

It was precisely this paradox that Schrödinger intended. Indeed, he forced physicists to position themselves more clearly with regard to the Heisenberg cut. Hardly anyone doubted that the quantum laws were valid in the microworld. Their calculations and predictions matched the measurements too closely. But the question was: how far does their power extend into the macro world? Where in the chain *atomic nucleus–measuring device–prussic acid bottle–cat–observer* must one place the transition from the quantum world to a macro world in which we can trust our usual views and objects to have well defined and independent properties?

- Born and Pauli assumed the Heisenberg cut would occur directly after the quantum object, thus transferred to Schrödinger's thought experiment, between the atomic nucleus and the Geiger counter. For them, there was no cat in an indeterminate state; as a component of the macro world, it had to be objectively either dead or alive.
- Bohr saw the cut in the first macroscopic measuring component, in the Geiger counter, since the processes that cause the Geiger counter to deflect clearly take place in the microworld and are subject to quantum laws.
- John von Neumann, who will be discussed in a later chapter, assumed the transition to occur between the measuring apparatus and the observer.
- Heisenberg believed that the exact position of the dividing line could be chosen *arbitrarily*.[1] He relied solely on the mathematical description of the experiments, and this was independent of the position of the cut. How a person chose to imagine the quantum world had no meaning for him.

More than 25 years later, the American physicist **Eugene Wigner** introduced another link into the above chain of atomic nucleus, Geiger counter, poison bottle, cat, and observer: he himself observes from the door as his friend opens the box with the cat. From Wigner's point of view, the cat

---

[1] W. Heisenberg, *Wandlungen der Grundlagen der exakten Naturwissenschaft in jüngster Zeit*, lecture to the Society of German Natural Scientists and Physicists, Hanover, 17 September 1934, Angewandte Chemie 47 (1934).

continues to be in a superposition of "dead" and "alive" until his friend—degraded to a "measuring device" in this experimental setup—calls out to tell him what he sees in the box. Only then does the superposition of indeterminate properties become an unambiguous reality. For Wigner, then, only the consciousness of the observer represents the boundary between quantum mechanics and classical mechanics. He wrote:

> It was not possible to formulate the laws of quantum mechanics in a perfectly consistent way without reference to consciousness.[2]

This thought experiment, which became known as "Wigner's friend", was interpreted by him as an indication that human consciousness is the property-determining function for quantum systems. But now another paradox comes into play: there could also be third, fourth, and fifth observers, all connected in series, so to speak. Depending on which of the observers is classified as relevant, the quantum mechanical state reduction takes place at different points in the chain. It is therefore itself highly subjective. In the end, this seemed too abstruse to Wigner himself, and in the 1970s he discarded his idea of a reality that was first brought about by human subjective consciousness. The problem of the Heisenberg cut thus remained open.

A New Addition to the Vocabulary of Philosophers and Physicists

And Schrödinger himself? Where did he make the cut? He found his own way of approaching this question: he looked at the measurement process, because that is precisely where a device from the macro world comes into contact with a quantum object.

Only the wave function of the single electron orbiting the nucleus in the hydrogen atom can be fully calculated. Many-particle quantum systems have to be described by a single wave function which is an element of the infinite-dimensional Hilbert space. This is already so complicated for the two electrons in the helium atom that mathematics reaches its limits. For all systems that go beyond this—from the lithium atom, with three electrons, to the entire universe—mathematicians have to work with simplifications. These work so

---

[2] E. Wigner, *Remarks on the Mind–Body Question*, Mehra, J (ed.), *Philosophical Reflections and Syntheses*, Springer, Heidelberg, (1983), pp. 247–260 (Original in: I. J. Good (ed.), *The Scientist Speculates*, Heinemann & Basic Books, New York, pp. 284–302 (1962).

well, however, that even larger quantum systems can be calculated extremely precisely.

Physicists already knew that a system of several quantum objects cannot be described as a combination of separate one-particle states, but only by a *common* wave function Ψ. Even in the case of the helium atom with its two electrons, it was known that a common wave function had to be calculated; it could not be treated as some sort of combination of two individual wave functions, one for each electron. While other physicists strictly separated the macroscopic world from the quantum world, Schrödinger consequently applied the principle of the common wave function to measurement as well, i.e., to the moment when the macroscopic measuring system comes into contact with the quantum entity to be measured. Somewhat casually, he remarked:

> [The Ψ-function of the measured object] has, according to the inevitable law of the total Ψ-function, become entangled with that of the measuring instrument [...]

Schrödinger thus introduced the word "entanglement", which is still used today to describe this fundamental idea. In 1935, the same year as he had presented his thought experiment with the cat to the public, he wrote in another article:

> This property [entanglement] is not **one**, but **the** property of quantum mechanics, the one in which the whole deviation from the classical way of thinking manifests itself.[3]

Entanglement is the reason why a quantum particle can never be considered on its own. Anyone who wants to measure an electron, for example, must accept that it forms a common quantum system with the measuring instrument used for this purpose. This realisation has fundamental implications for physics:

- Each system would be connected to the next-higher measuring system (Geiger counter, poison bottle, cat, observer, etc.) in an ever-increasing wave function.

---

[3] E. Schrödinger, Discussion of Probability Relations between Separated Systems. *Proceedings of the Cambridge Physical Society*, 31 (1935), p. 555.

- In the final analysis, the entire universe would have to be included in the total wave function for every measurement.
- There would be no Heisenberg cut; the quantum laws would also apply in the macro world. The cat would indeed be in a state of superposition as long as no one looked to see if it was alive or dead.

> **Entanglement:** Schrödinger introduced this term, which still characterises the discussion about quantum physics, in the same year in which he presented his thought experiment with the cat.

These consequences diametrically contradict our experiences, because somewhere on the scale of magnitude, the indeterminacy of the quantum world is quite obviously replaced by the determinacy of the macrocosm to which we humans are accustomed and adapted.

The exact location of the Heisenberg cut is still the subject of research today. In recent decades, increasingly clear evidence has emerged that the boundary is by no means absolute, and that quantum effects such as Bose–Einstein condensation, laser properties, superconductivity, and superfluidity can indeed occur in the macroworld.

Einstein Provides a Solution for Himself

The most bitter opponent of the Copenhagen interpretation was Albert Einstein. Of course, he recognised and used the possibilities of quantum theory to calculate atomic phenomena in a mathematically accurate way. But for philosophical reasons he could not and would not accept certain points:

- That the atomic world should not be directly experienceable for us and should be representable solely through mathematics.
- That quantum particles have no objective properties independently of measurements and therefore have neither **substance nor identity**, and thus strictly speaking, no **reality**.
- That the **causality** so wonderfully captured in laws by Newton should not apply in the quantum world. The Schrödinger equation can be perfectly interpreted statistically, but the fuzziness of location and momentum means that the system can no longer be described exactly deterministically, and only statistical statements are possible. He did not agree with the view

that, in Schrödinger's cat experiment, it should not be possible to calculate whether one of the atoms decays and sets the chain of events in motion or not.

- That instead of clear quantum states, such as *spin up* and *spin down,* there should only be indeterminate intermediate states. In his opinion, this could not be, if only because this would mean that another basic assumption from classical philosophy would no longer be valid: "*Tertium non datur*"— the law of excluded middle.

There was another point that troubled Einstein: Bohr's interpretation presupposed that an instantaneous transfer of information takes place when the wave function collapses. If, for example, the wave function of the electron collapses in the double-slit experiment, the information that the electron can no longer hit at a given point is transmitted to all other places *without any time delay*, i.e. at infinite speed. This instantaneous decay of the wave function postulated by the Copenhagen group was the centre of discussion at the Solvay Conference of 1927, the fifth time after 1911 that the leading physicists from all over the world had met in Brussels at the invitation of the industrialist Ernest Solvay. It was on this occasion that the famous Bohr–Einstein debate took place. The great scientific contributions of these giants of physics had already been made some years before, so the philosophical debate between them was all the more heated:

- Einstein argued that, according to his theory of relativity, the transmission of matter or information faster than light was impossible. Therefore, the instantaneous collapse of the wave function was impossible; he called this long-distance effect "spooky".
- Bohr replied that no remote effect was responsible for the collapse of the wave function, but that this simply corresponded to the character of the wave function as a probability wave. According to the uncertainty principle, there is a fundamental uncertainty about the exact location of a single quantum particle at a certain point in time. This means that the exact time at which something happens to a quantum object is fuzzy anyway, so it cannot be specified exactly. The concept of "instantaneous information transfer" would only make sense if the whereabouts of the quantum objects could be clearly defined.

Bohr's argument did not convince Einstein. To the end of his life, he had no doubt that fundamental components of quantum theory were still missing. In 1935, these so-called "hidden variables" had already been haunting the minds of some physicists for several years, including Wolfgang Pauli and Louis de Broglie, who had in fact presented a deterministic hidden variable theory in 1927, but then discarded it after lengthy discussions with Wolfgang Pauli. Einstein took up this idea again: hidden variables ensure that the properties of quantum objects recorded during a measurement are already fixed before the measurement, even if they themselves cannot be measured. These objective properties would make incomplete quantum mechanics a consistent theory compatible with classical physics, in which ...

- ... there is no superposition,
- ... an objective reality exists, and
- ... events proceed deterministically.

In 1932, **John von Neumann** published a mathematical proof that the variables Einstein was looking for could not exist.[4] As early as 1935, the philosopher and physicist Grete Hermann[5] knew that this proof was flawed because he had "illegally" transferred a link in his chain of proof from classical physics to quantum physics. Since she was in tensive exchange with the leading German physicists—Heisenberg even mentions Grete Hermann at length in his autobiography—they must have been aware of her easily comprehensible objection. It is quite incomprehensible that it should have been ignored by the physics community.

Von Neumann's false proof continued to hold sway into the 1960s and in all probability prevented faster progress in quantum physics. In 1966, John Bell pointed out that von Neumann's assumptions did not always apply.

In 1935, Albert Einstein and his American assistants Boris Podolsky and Nathan Rosen published an article entitled "Can quantum–mechanical description of physical reality be considered complete?"[4] The thought experiment described therein, later referred to as the Einstein–Podolsky–Rosen paradox (or EPR paradox, for short), was as ingenious as it was

---

[4] In von Neumann's book *Die mathematischen Grundlagen der Quantenmechanik* (Springer 1932), the proof appears in Chapter VI.

[5] G. Hermann, *Die naturphilosophischen Grundlagen der Quantenmechanik*, §7, Abhandlungen der Fries'schen Schule (ASFNF), vol. 6, issue 2 (1935), p. 99.

[4] A. Einstein, B. Podolsky, and N. Rosen, *Can Quantum–Mechanical Description of Physical Reality Be Considered Complete?* Phys. Rev. 47, 777 (15 May 1935).

simple. Two quantum particles in superposition with each other are parts of a common wave function. Their momentum cannot be determined exactly, but it is known that the momentum of one particle is directly related to the momentum of the other particle. This is what the Copenhagen interpretation says. Now the two particles are separated from each other and brought to two different places. If a measurement of the momentum is then carried out on one of the particles, the following must happen:

- The wave function of the measured particle collapses; a fixed momentum quantity can now be assigned to it.
- Thus, the common wave function also decays. The properties of the partner particle are now also well defined, and that includes its momentum, because according to the law of conservation of momentum, this value depends directly on the measured value of the other particle (today, the EPR paradox is mostly explained using the example of two electrons and their spins).

With this thought experiment, called the EPR paradox, Einstein, Podolsky, and Rosen wanted to draw attention to the abstruseness of superposition. It was a very clever move, which Bohr had little to oppose. Nevertheless, the shot backfired, albeit only later, when Einstein and Bohr had both passed away: it is precisely the ridiculed and seemingly impossible instantaneous action at a distance that is the basis of many technical applications today, not least a possible future quantum computer.[5]

In 1927, the Bohr–Einstein debate at the Solvay Congress had been a joint struggle among physicists over central philosophical questions. In 1935, the dispute about the philosophical background of the Copenhagen interpretation had reached its climax; since Einstein and other researchers such as John von Neumann had already emigrated to the USA, it sometimes took place over great distances. But after that, the discussion abruptly broke off; hardly anyone was interested in the inner contradictions that had caused so much disagreement to Einstein and Bohr. The generational change was complete: The vast majority of physicists were no longer bothered by the fact that they could not explain why quantum mechanics worked mathematically and in experiments—they simply applied it. (The following chapter reports on another decisive reason.)

Only the four greatest physicists of the twentieth century—Einstein and Schrödinger as critics of quantum mechanics, Bohr and Heisenberg as its

---

[5] For more examples and elaboration on this, see: L. Jaeger, *The Second Quantum Revolution,* Springer (2018).

advocates—continued the search for an explanation to the very end. Einstein died in 1955, Schrödinger in 1961, Bohr in 1962, and Heisenberg in 1976. Despite all their efforts, they had to leave open the questions about spooky remote effects, entangled particles, and half-dead cats. For a long time, they were the last physicists for whom the philosophical background of their subject had any relevance. For the others, the successes of mathematics had become more important than unanswered philosophical questions.

So it was that, for a period of about 30 years, the EPR paradox and Schrödinger's cat disappeared from the focus of physicists. But then something very surprising happened. In the 1960s, the Irish physicist John Bell succeeded in setting up an inequality with the help of which it became possible to verify *by experiment* whether the hidden variables predicted by Einstein existed or not. It became the prelude to a new approach to quantum physics—ignited by the phenomenon of entanglement. With this, it became clear over time that ...

- ... there is no boundary between the microcosm and the macrocosm,
- ... the bizarre properties of quantum physics are not limited to the microcosm,
- ... macroscopic quantum physics enables entirely new technologies,
- ... human consciousness plays no role in the quantum world.

Mathematics Becomes Substance

Most physicists no longer ask philosophical questions. Nor do they search for the "true existence and ultimate essence of things", as philosophers since the pre-Socratics, but also physicists like Galileo, Newton, and even Einstein had done. Quantum physics had dissolved the concept of substance (and thus also of reality, identity, and causality).

The mathematisation of everyday objects is not an invention of quantum physics. We find it already with Pythagoras, and later with Archimedes. Copernicus, Kepler, Galileo, and, most recently, Newton, did nothing other than translate the phenomena of nature into the language of mathematics. But there is a big difference between the mathematics of classical physics and that of quantum physics:

- In classical physics, the truth of mathematically formulated statements about the world in the reality we perceive can be directly verified by experiment and observation—planetary motions in the night sky are easy to follow.
- The bizarre phenomena and apparent paradoxes of quantum physics cannot be observed directly. Because they take place beyond our horizon of experience, we cannot even put them into words in our everyday language. Only by means of abstract mathematical descriptions are physicists able to grasp the processes in the atomic world, to formulate their theories consistently and, in this sense, to "understand" nature.

It was a major step in human history to replace the ideas and perceptions adapted to the scales of magnitude familiar to man with a purely mathematics-based, "insubstantial" view of the world. This literally world-shaking development took place far from public perception. At the beginning of the twentieth century, the fathers of quantum physics could perhaps consider themselves lucky that their theory was too complicated to call the ecclesiastical authorities to the scene as it had in Galileo's time (even if they had lost their claim to absolute interpretive sovereignty, they still possessed some influence). For one thing is clear: the rejection of any independent substance—and thus also of a single, eternal truth—represents a far more serious attack on religious dogmas than Galileo's teaching on the motions of the heavenly bodies.

Surprisingly, the old idea of an unchanging substance has recently crept back into today's physics. For the majority of theoretical physicists, mathematics is no longer just a tool to describe and understand the world, but what philosophers call a "perfect idea": fundamental and valid everywhere and always. In modern philosophical interpretations of quantum theory, mathematical structures and concepts, such as symmetries, conservation laws, and invariances, are elevated to the rank of substances. In the spirit of Plato, mathematics thus becomes the independent basis of everything. As the bearer of absolute properties, it would still be there even if there were no more people to conceive of it, and even if there were no more things to be counted and related to each other.

With the idea of mathematics as an absolute quantity, the metaphysical belief in something fundamental plays a role again today. The physicist and philosopher Carl-Friedrich von Weizsäcker put it this way:

And therefore, if one asks why mathematical laws apply in nature, the answer is: because these are its essence, because mathematics expresses the essence of nature.[6]

The Current State of Philosophy

Mathematics can therefore be regarded as immaterial substance. And what about material substance? Quantum objects have no physical properties. Mass, charge, and spin are qualities that come into play solely through interaction with their environment. Philosophically, this means that there is no substance in the classical sense, but only accidents in the form of transient interactions between quantum objects. This results in a fundamental difference between the macro world and the quantum world:

- In the traditional metaphysics of Western philosophy, interactions come into the world as a consequence of existing things.
- In the quantum world it is the other way round: interactions constitute things.

This concept of reality is not entirely new and unique in the history of Western thought:

- In the teachings of some of the pre-Socratics, the dualism between substance and accidents had not yet developed its philosophical dominance.
- Kant's philosophy already echoes the concept of reality in quantum physics: objects only acquire their fundamental properties through our perception and our thinking.
- The philosopher Edmund Husserl also speaks of two separate dimensions in the process of cognition: the "stream of consciousness", i.e., the way information is processed in our brain, and the phenomenon towards which it is directed.

Some physicists and philosophers go in a completely different direction. Given the results of the double-slit experiment, in which it depends on the

---

[6] C. F. von Weizsäcker, *Ein Blick auf Platon - Ideenlehre, Logik und Physik,* Stuttgart 1981.

observer whether the electron behaves as a wave or as a particle, they advocate a completely subjectivist interpretation of the quantum world[7]:

- Quantum objects are products of conscious observation (this is where "Wigner's friend" comes into play again).
- Therefore, even in the macro world, everything observed exists solely through an observer—the tree is only in the forest when it is looked at by a conscious observer.
- All experiences depend exclusively on the relevant observer.
- All knowledge is exclusively subjective.

The adherents of this worldview, the so-called subjectivists, not only reject the substantial existence of quantum particles, but deny them *any reality at all.* Since there is no objective truth about them, nothing in the world is real, not even quantum particles. Subject-independent truths are fundamentally excluded. Thus, for the subjectivists, neither classical physics nor our everyday experiences are anchor points to which we could adhere. Consciousness alone constitutes existence (which, however, is neither true nor real).

Reality is that which exists without illusion and not solely according to the perception of an individual person. It is therefore not dependent on how an observer perceives it, and it is not dependent on there being any observer at all.

Truth can be a reality, but also a subjective perception that is assessed as "true". For subjectivists, truth is *always* subjective.

Presumably, in the imagination of the subjectivists, conscious perception plays an all too absolute role. For from the fact that independent particles with their own identity do not exist in the quantum world, we cannot necessarily conclude that there are no particles and *no objects at all*! The process of measurement does not *create* quantum objects, it only brings them into a state that is observable for us, for example, that of a wave or a particle. The quantum world can therefore exist in reality, just not with objectively defined properties.

---

[7] For a modern interpretation see C. Caves, C. Fuchs, R. Schack, *Quantum probabilities as Bayesian probabilities*, *Phys. Rev. A* 65 (2002), 022305.

Physics and philosophy have moved a little closer together these days. One can expect their interplay to yield further exciting insights into what builds and holds our world together.

8

The Final Dissolution of All Matter—The Shift from German to American Physics

Until the 1930s, the development of quantum mechanics and relativity theory was largely driven by German-speaking scientists. Göttingen, Zurich, and Berlin were the centres of theoretical physics and mathematics worldwide, and German was also spoken predominantly in Niels Bohr's Copenhagen group. Although Danish, Bohr himself spoke and loved this language like his mother tongue. Even the Englishman Paul Dirac spoke fluent German.

- Max Planck developed his quantum hypothesis, the big bang of quantum theories, so to speak, at Berlin's Friedrich Wilhelm University in 1900.
- Albert Einstein's photon theory was developed in Zurich in 1905.
- Werner Heisenberg in Göttingen and the Austrian Erwin Schrödinger, who was living in Zurich at the time, developed the first quantum theories in 1925.
- Other important researchers in the quantum world were Wolfgang Pauli, born in Vienna, whose path took him to Munich, Göttingen, Copenhagen, and finally to Zurich, and Pascual Jordan, who came from Hanover and worked in Hanover, Göttingen, and Rostock.

The development of the mathematical foundations of the new physics was also shaped by German-speaking mathematicians: David Hilbert, Emmy Noether (about whom we will read in detail in the following chapter), and Hermann Weyl worked in Göttingen. Weyl, who was born in Germany and

© The Author(s), under exclusive license to Springer Nature Switzerland AG 2022
L. Jaeger, *The Stumbling Progress of 20th Century Science*,
https://doi.org/10.1007/978-3-031-09618-1_8

was in Zurich at the same time as Einstein, further developed the mathematical foundations of the theory of relativity and took over Hilbert's chair in Göttingen in 1930.

From 1901 to 1935, 39 of 111 **Nobel Prizes in Science** went to German-speaking researchers, more than one in three. In contrast, there were only two Nobel Peace Prizes for German-speaking laureates during this period.

The best students and young scientists in physics and mathematics flocked to Germany from all over the world to learn and do research here. How did this preponderance of German-speaking physicists and mathematicians come about? In the early 1970s, the American historian of science Paul Forman provided an explanation that may be exaggerated in many respects, but probably contains a grain of truth.[1] According to his findings, the remarkable progress in the early development phase of quantum physics was related to the prevailing zeitgeist in Germany at the time.

Before the First World War, classical physics was still firmly bound to rationality, causality, and reality; everything had to be calculable, derivable, and comprehensible. Ideas that pointed beyond these limits hardly had any chance of being heard—the fact that the brilliant mathematician Georg Cantor had to make a living in the provinces because he was working with different kinds of infinity is an example of this rigid view of science. Even the physicist Ludwig Boltzmann, who introduced random calculations and statistics into physics, was initially met with vehement rejection. The shock of the First World War knocked a breach in this wall which had hitherto locked scientific thinking into this dead end. Germany had lost the war, the Empire had collapsed; the 1920s and early 1930s were marked by great hostility toward the newly formed Weimar Republic, intellectual circles being no exception. Out of the economic depression and social instability developed an intellectual revolt that broke with everything that had been and was generally directed against the determinism and materialism of the pre-war period. In its place came an almost romantic attitude of intuition and irrationalism. Forman suggested that young German-speaking physicists were influenced by this social climate and therefore dared to take the step of detaching the previously untouchable causalities of Newton and Maxwell from quantum mechanics and developing completely new ways of thinking.

---

[1] P. Foreman, *Weimar Culture, Causality, and Quantum Theory, 1918–1927: Adaptation by German Physicists and Mathematicians to a Hostile Intellectual Environment*, Historical Studies in the Physical Sciences, Vol. 3 (1971), pp. 1–115.

Forman's thesis triggered an intense debate among historians. Even today, there is no consensus as to why the first advances into the quantum world were made in the German-speaking countries in particular. On the other hand, it is clear why German dominance in physics and mathematics ended. As early as 1930, when he took over Hilbert's chair in Göttingen, the mathematician Hermann Weyl noticed a change in mood:

> It is only with some trepidation that I find myself back from ... (the) freer and more relaxed atmosphere (of Switzerland) into the yawning, gloomy, and tense Germany of the present.[2]

This was followed by the Nazi takeover in 1933 and the ban on supposedly "Jewish" physics and mathematics. Many outstanding scientists had to leave Germany, mainly for the USA. Among those expelled were already important names, but also some highly talented young researchers:

- Albert Einstein was German, but hated the conservative atmosphere in the country of his birth. As a young man he was stateless for a few years and in 1901 he took Swiss citizenship, which he kept until the end of his life. Because he accepted a professorship in Berlin in 1914, he had to apply for a German passport again. In 1933, he gave up this citizenship for good and left Germany with the firm intention of never returning. In 1938 he became an American and now had dual citizenship once more.
- The Hungarian John von Neumann (who will also be mentioned later) had attended the German-language grammar school in Budapest and spent the 1920s and early 1930s in Berlin, Zurich, and Göttingen, among other places. Like Einstein, he emigrated to the USA in 1933.
- The New Yorker Robert Oppenheimer studied and taught in Germany. In the USA, he was later instrumental in the construction of the American atomic bomb.
- The Hungarian Edward Teller also emigrated to America and later played a significant role in the development of the American hydrogen bomb.
- Born in Strasbourg, Hans Bethe, who researched the electron in the 1920s and early 1930s, emigrated first to England in 1933, but later moved on to the USA.

---

[2] Hermann Weyl: *Gesammelte Abhandlungen*, Vol. IV, pp. 651–654, Springer-Verlag (1968); also available (in German original) at: https://www.spektrum.de/wissen/hermann-weyl-1885-1955-mathematischer-universalgelehrter/1371662.

The *brain drain* was extensive and left behind a devastated scientific landscape. When the mathematician David Hilbert attended a banquet in 1934, where he sat next to the Nazi Minister of Culture Bernhard Rust, the latter asked whether "the Mathematical Institute had really suffered so much because of the departure of the Jews". Hilbert's answer was: "Jelitten? It has not suffered, Minister. There is no such thing any longer!"

Within a few years, the German-speaking countries had little to say in the *theoretical* sciences, which they had dominated for so long. *Applied* physics and mathematics, on the other hand, made important and remarkable progress even during the Nazi era, because the Nazis knew about the importance of these disciplines for their war machine. Science now essentially took place outside the universities, for example in separate research groups attached to military units. In particular, scientists were supported by the Luftwaffe.

The assessment of Germany's political leadership that theory could be neglected in favour of practice proved fatal. The consequences were not only devastating at the level of individual destinies; without fundamental research and without abstract physics and mathematics, progress in technological applications stalled. The exodus of a substantial part of the scientific world soon began to take effect: From the 1940s onwards, the USA took the lead, not only in world politics but also in the natural sciences. Physicists or mathematicians who wanted to be on the cutting edge no longer went to Berlin, Göttingen, or Copenhagen, but crossed the Atlantic.

The shift of scientific activity from the German-speaking countries to America was more than a mere change of location: because a completely different intellectual climate prevailed in the USA, a new kind of theoretical and experimental physics developed.

- The time for philosophical discussions in physics came abruptly to an end. In America, people were not interested in the question of "what holds the world together at its core". It was supposed that mathematics would take over the explanation of the world, and that with its help, all the problems of physics would soon be solved.
- It was now a question of the technological application of scientific knowledge, not least for military purposes. While in Germany they tried to achieve the same end without any "diversions" via theory, the Americans understood that technology had to have a theoretical foundation to be successful.

With these objectives, it became possible in the USA for mathematics, which had already lost its concrete descriptiveness in Europe, to become increasingly abstract and complex.

The Electromagnetic Field is Quantised

What was the state of quantum research in the mid-1930s?

- **Quantum mechanics (QM)**: Two fundamentally different quantum theories existed for quantum objects such as electrons. Both described *how particles can also behave as waves*. The first theory of quantum mechanics was based on the work of Schrödinger and Heisenberg, whose findings had been shown to be perfectly analogous. Their approaches were non-relativistic, i.e., they still operated within the classical concepts of space and time. The second theory was Paul Dirac's, which was relativistic because it took into account space–time as described by Einstein.
- **Quantum electrodynamics (QED)**: In order to fully understand and describe the wave–particle duality, the opposite path also had to be taken: How can it be that *waves can behave like particles*? This derivation was not possible from QM. We were still completely in the dark as to how this path could be derived mathematically. The only known theory was Maxwell's classical electromagnetic field theory, which describes the wave character of free electromagnetic fields. The spectacular discovery in 1932 of the anti-electron (positron) predicted by Dirac encouraged physicists to expand Maxwell's field theory into a quantum theory.

In 1927, one year before the publication of his famous equation, Paul Dirac had already sketched out the first steps towards a mathematical description of quantum fields.[3] In this "quantisation of the field", he had attempted to represent the classical field equations in such a way that the physical measurands of electromagnetic waves could only assume quantised values. At first, his theory was non-relativistic, i.e., it still operated with separate concepts of space and time. Later, it became apparent that the Dirac equation could be interpreted in such a way as to take into account the relativistic nature of the electromagnetic field, i.e., the space–time described by Einstein.

---

[3] P. Dirac, *The quantum theory of the emission and absorption of radiation*, Proc. R. Soc. London A 114, 243–265 (1927).

What turns electromagnetic waves into photons? This was precisely Planck's and Einstein's question at the beginning of the twentieth century, and it gave rise to quantum mechanics, which describes the opposite situation: how particles become waves. Physicists therefore started all over again, where Planck and Einstein had begun thirty years earlier.

Despite this progress, QED was not yet satisfactory, if only because it did not provide an answer to experimental observations.

- **The problem of free electrons**. Many experiments, including the double-slit experiment, used electrons fired into experimental setups. While electrons bound in atoms can only absorb and release energy packets of a certain size, free electrons can assume any energy states.
- Since the early days of quantum physics, it had been known that **particles can disappear in experiments or appear as if from nowhere**. The best known example of such a process is the absorption or emission of photons in the atom. In 1932, Irène Curie and Frédéric Joliot had experimentally demonstrated another "magic trick" of nature (without, however, grasping the significance of this observation): electron–positron pairs can emerge from a high-energy photon and disappear again immediately afterwards. Einstein had described such processes mathematically and Bohr had prepared the equations statistically. But no one could explain what was actually happening.

The fact Dirac's QED equations could not describe these two phenomena was partly because they worked with constant particle numbers. Although they *predicted* the creation of spontaneously arising electron–positron pairs and their immediate subsequent decay, they did not seem to be able to represent suddenly appearing or disappearing quantum objects. It was only later realised that Dirac's equations could indeed be equipped with further interpretations in such a way that particle numbers could change.

Dirac was one of the few researchers who continued to conduct research in Europe after the mid-1930s and enrich quantum physics with relevant findings. This could be due, among other things, to the fact that he had already returned to England from Germany in 1932. When Hitler came to power and the persecution of the Jews began, he took a vow never to speak German again. In Cambridge, he taught in the Department of Mathematics—as the thirteenth successor to Isaac Newton.

The Photon Drops Its Mask

The fact that photons are energy carriers that load "packaged" amounts of energy from one quantum object onto the next was already known around 1930. It was obvious that the electromagnetic field also obeys quantum rules. Physicists' attention now turned to the interactions between electrons and photons. There was a picture of their interaction that allowed at least a rough idea of the processes. An electron is thrown onto a stretchable surface, a kind of stretched rubber blanket. It sticks to the rubber blanket without resonating and causes it to bulge downwards; it passes on all its energy to the surface in the form of a wave. The wave passing over the blanket hits a second electron, which is thrown upwards. In the quantum world, this energy transfer happens suddenly and completely. Just as completely as the first electron gives up its energy, the second electron takes up the energy.

- In Bohr's atomic model, the electron excited by a photon jumps to the next higher **orbit** around the atomic nucleus.
- In the illustration of quantum mechanics, the energy of the photon stimulates the electron to jump from one **state** to another.

Field: the spatial distribution of forces acting on electric charges.

Electromagnetic field: since electric and magnetic fields cannot be separated from each other, the two are usually combined into *one* phenomenon.

Waves: a field that propagates in space by periodic oscillations perpendicular to the direction of propagation.

What exactly this "state" is and whether the electron must now be seen as a particle or a wave was still hotly debated. In addition, there was still the problem that the photon comes into being "out of nothing" and then disappears back into nothing again. How could one represent mathematically where it comes from and where it stays when it has given up its energy? In order to solve these puzzles, new, incredibly abstract ways of calculation had to be worked out—and the USA was exactly the right breeding ground for this.

When the mathematicians and physicists further developed and reinterpreted the already known equations, they were astonished to discover that light could no longer be represented as a wave, but only as a particle! What a *twist in the plot*! For centuries, people had been sure that light was a wave. When Einstein postulated in 1905 that light waves could also have a particle nature (in the form of photons), it was a revolution. Twenty-five years later, people came to the conclusion that photons definitely do not occur as waves, not even in the form of a wave–particle duality, but as energy carriers that transport energy packets from one quantum object to the next. Since the wave–particle duality was still assumed, photons were now called "exchange particles". In fact, photons exhibit certain particle properties: as long as they do not interact, they do not decay. Emitted by distant stars, they can travel billions of years through space before they hit the earth. But they also have properties that definitely do not fit a particle, for example they are non-local and can behave like a wave at a double slit. Heisenberg's uncertainty principle had already shown that the location of quantum particles, including photons, is "blurred".

Little by little, the veil was lifted: the photon was *neither a wave nor a particle*. Nor did it carry an energy packet around with it. In fact, it *is* the energy packet! One had to say goodbye to the idea of waves and particles and get used to speaking only of different energy states of the electromagnetic field.

- In the theory and equations of **quantum electrodynamics**, a photon is an exchange particle which, viewed graphically as a small sphere, absorbs an energy packet from a larger sphere, an electron, and emits it to another large sphere.
- Now a new concept emerged, which, after quantum mechanics and quantum electrodynamics, found its representation in a third theory, **quantum field theory**: everything happens on the basis of an all-encompassing electromagnetic field. Photons are energetic parts of this field. They are generated by energetic electrons (which lose energy in the process) and reabsorbed by other electrons (which gain the energy). This exchange transfers electromagnetic force from one electron to the other. There is no transfer of mass because the photon travels at the speed of light and therefore cannot have any mass.

Even theoretical physicists who turn humanity's view of the world on its head can be very conservative. Although they had established that photons are pure energy, the name "exchange particle" stuck. And depending on the field of research, there is still talk of particles and waves, even though it is clear that all these quantum entities are nothing more than energetically excited electromagnetic fields.

> Photons are **exchange particles** for energy. They are involved in every process in which energy is transferred. Physicists today believe that there are special exchange particles for *all* physically acting forces. The exchange particle of the strong nuclear force, for example, is the gluon. However, the exchange particle responsible for gravity has not yet been discovered.

Fluctuations and More Permanent Emanations of Energy

With the detection of photons as the electromagnetic exchange particles, two decisive steps had been taken:

- Now it could be theoretically recorded and mathematically proven that field and "particle" can each emerge separately, since they are in principle one and the same.
- It was now known why electrically charged particles attract or repel each other: electromagnetic energy is transferred by photons as exchange particles.

The new quantum field theory proved to be much more complex than the quantum electrodynamics of the 1930s. Thus, electrons and photons, but also all other quantum entities, can exist in two different states due to their quantum nature: "real" or "virtual":

- Quantum entities are in a **virtual state** when they occur as spontaneous fluctuations of the electromagnetic field. Although they cannot be measured directly, their effect on their environment can be detected. The basis for the fact that electron–positron pairs, for example, can appear spontaneously (and immediately disappear again) without external energy supply and even in a vacuum is made possible by the Heisenberg uncertainty principle: just like the product of location and momentum, the product of energy and time also has a fuzzy value. This means that, at any given place in the world, its energy content cannot be *zero*. However, this also means that we cannot rule out the possibility that a quantum object will appear at this location for a short time as a spontaneous fluctuation of the quantum field.

- In the **real state**, particles are stable and directly measurable. If sufficient energy is supplied, the resulting quantum objects can be long-lived. Electron–positron pairs have the easiest time "jumping into reality", because they have a low mass compared to protons or neutrons and require relatively little energy for their creation. Other particle–antiparticle pairs need correspondingly higher energy inputs. Everything we perceive in the macrocosm is made up of real particles. They have absorbed so much energy that they are directly measurable beyond the background noise of the spontaneous fluctuations of the electromagnetic field.

Virtual particles are created when a fluctuation around the zero value occurs in an electromagnetic field.

They only exist for a very short time and immediately disappear into their environment. Virtual particles include photons that are bound in an atomic system and transfer energies there.

Real particles have enough energy to exist in real terms—energy becomes real particles. If a photon changes into a real, free state by absorbing energy, it can be detected as a beam.

In 1905, with the formula $E = mc^2$, Einstein had explicitly shown for the first time in his special theory of relativity that energy and mass are equivalent. More than twenty-five years later, this connection reappeared in the new quantum field theory: quantum particles with mass decay into energy and can also spontaneously emerge from it. It took almost another decade before the equivalence of energy and mass was also proven experimentally. In December

1938, Otto Hahn and Lise Meitner showed by splitting heavy atomic nuclei that a part of their masses was converted into energy. This was the last work of great significance for quantum physics to be carried out in Germany. Lise Meitner left her home country only a few days after successfully splitting the atom for the first time; following a stopover in Scandinavia, she also found herself in the USA. There, atomic fission became the basis of atomic bomb technology. Less than seven years later, the first three were detonated in New Mexico, Hiroshima, and Nagasaki (see Chap. 9).

In quantum physics, **mass** is no longer bound to substantial particle properties but, via $E = mc^2$, to a special property of high-energy electromagnetic fields and, as was soon recognised, other force fields as well.

The Disappearance of Matter

Is it possible to somehow imagine the construction of our world from quantum objects? Unfortunately, any possibility of "concrete" visualisation disappears in quantum field theory. Even electron balls falling on rubber mats can no longer keep up.

- A quantum object is not a small piece of matter, as physicists from Democritus to Rutherford to Einstein had imagined. It is rather a collection of clouds of energy fields and exchange particles that interact with each other and cannot be considered separately.
- It is permanently surrounded by fluctuating quantum fields and thus by clouds of virtual particles that are constantly emitted and absorbed and interact with other quantum objects as exchange particles. Both the exchange particles and the interactions with other particles are essential components of the quantum objects themselves.
- Even a vacuum is only an *apparently* empty space. In it, virtual particles can emerge as if from nowhere—and with the appropriate energy supply, real, observable particles as well, including their interactions with each other. The latter are in turn connected to virtual particles.

If quantum mechanics had already proved difficult to understand by around 1930, quantum field theory required far more abstract forms of mathematics. The example of trajectories (these are the solutions of systems of differential equations, also called orbitals) shows how far the new, "American" mathematics had broken away from classical physics and how complex everything had become.

- In **classical physics**, bodies act directly on each other—a tennis racket hits the ball and sends it on its journey through direct energy transfer. The trajectory results in a path on which the tennis ball can be assigned a clearly defined location at any point in time.
- In **quantum mechanics**, things became more complicated. The idea of Bohr's atomic model with its well-defined electron orbits around the atomic nucleus had already been called into question by Bohr himself at the beginning of the 1910s. By the mid-1920s, the work of Schrödinger and Heisenberg had led to this idea being shelved. It was now known that quantum objects had to be described by wave functions whose solutions are broadly scattered trajectories. Instead of a line on which the particle moves, there are spaces in which it can be found according to statistical rules— some readers may still remember the orbital models from school lessons. The interaction between bodies is also different in the quantum world. In quantum mechanics, there are no individual particles with definite trajectories that interact directly with other particles. Rather, all particles form a common whole in which all components permanently interact with each other, each particle simultaneously with every other.
- In **quantum field theory**, interactions take place between the trajectories of different quantum particles. Strictly speaking, all quantum objects of the entire universe would even have to be taken into account, but in practice one limits oneself to the immediate surroundings of the quantum object under investigation. This is time-consuming enough and requires the calculation of extensive and complicated integrals with numerous variables. The probabilities of all transitions of states of all wave functions involved *before* the interaction must be related to their possible states *after* the interaction. Thus, chemists today who want to calculate the structure and reaction behaviour of large molecules need very powerful computers that cost many millions of dollars.

Today, **quantum electrodynamics** (QED) is part of the field of **quantum field theory**. Its core statement is that there is no material substance in the smallest particles that holds the world together. This task is performed by the *interactions* between them.

Classical physics and large parts of the Western philosophical tradition had relied on the belief that the solidity of atoms gives matter its stability. Bohr's atomic model had already made the first breach in this view in the early phase of quantum theory: with the electrons orbiting far away from the atomic nucleus, it was clear that 99.99% of matter consists of empty space. QED also eliminated the last 0.001% of "solid mass"; the classical idea of matter and of substantial integrity was now finally off the table.

The New American Quantum World

By the early 1940s, many of the German-speaking theoretical physicists who had so vigorously developed their subject in the Old World had long since emigrated to the USA. However, there were also brilliant American contributors to this science. One of them was the physicist Richard Feynman, born in 1918. He succeeded in developing a new interpretation of the Dirac equations so that it was now possible to calculate the electromagnetic interactions between quantum entities exactly. In order not to lose track of the individual calculation steps, he developed very abstract mathematical tools with which the complexity of the trajectories could be systematically recorded. Feynman had recognised that the integrals occurring in the wave functions had a certain regular structure and that each one could be traced back to certain mathematical building blocks. From this observation, Feynman derived an ingenious trick that is still used intensively today: for the theoretical description of the behaviour of the wave function of a quantum object during its interaction with other quantum objects, he introduced new mathematical operators called "propagators". Using these, he drew diagrams whose lines represent certain calculation rules for the propagators. In a sense, he created directions for mathematicians venturing into the otherwise impassable jungle of quantum field theory.

Of course, Feynman's propagators do not describe real paths of particles or locations of interaction and must not therefore be understood as a description of concrete spatiotemporal processes. The interaction between two quantum objects through exchange particles is a process that takes place in many dimensions and cannot be represented by human comprehension. Against this background, Feynman's diagrams work very well. This is shown, among other things, by the fact that a certain physical quantity, which had caused physicists immense problems for many years, could be determined very precisely from them. From quantum mechanics, it had not been possible to calculate the proportionality factor between the angular momentum and the magnetic moment of the electron, the so-called g-factor. The theoretical value, which results from quantum field theory and the use of Feynman diagrams, agrees with the experimental measurements to twelve decimal places. For no other value in physics or any other science is there such exact agreement between theoretical calculation and experimental measurements.

Feynman diagrams illustrate the very abstract computational processes for the description of quantum fields and thus help theoretical physicists not to lose track of computations involving complex total wave functions. The diagrams look simple and clear, but the mathematics behind them is extremely complex.

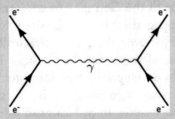

The diagram above describes the exchange of a virtual photon (γ) from one electron to another. Below is a higher order process.

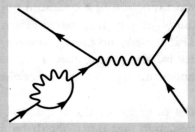

Feynman was one of the most brilliant physicists of the twentieth century, but he was also an exceptionally good communicator of his subject. To this day, his textbooks are very popular with students of theoretical physics. In addition, he was also a very amusing writer. His autobiography *Surely You're*

Joking, Mr. Feynman![4] is a very readable and extremely funny bestseller, even for non-physicists.

Modern quantum field theory from the USA was a triumph of very complex, but at the same time purely pragmatic mathematics. It required virtuosity in dealing with variables and operators rather than the ability to think through difficult conceptual problems or the philosophical questions associated with them. Put simply:

- In the European tradition, physics had begun with theoretical and philosophical concepts; a second step had been their translation into mathematics.
- The Americans started with mathematics and only then tried to find theories to go with it. The physicist David Mermin summed up this new methodological style in theoretical physics with the maxim: "*Shut up and calculate*".[5]

Perhaps it was precisely the focus on mathematical abstraction that enabled the new theories that emerged from it to achieve an agreement between theory and practice that had never before been contemplated in the history of science—in some cases, the deviation between experimental measurements and theoretical calculations (as for the g-factor above) was only 0.00000001%. By way of comparison, the orbit of the moon can only be calculated with approx. 99.3% accuracy using Newtonian physics; at an average distance of about 380,000 km, this corresponds to a difference of over 2600 km. If Einstein's theory of general relativity is taken into account, the difference between the calculated and measured position of the moon becomes much smaller; now there is less than a centimetre difference between observation and calculation. But even this value is nowhere near the accuracy of quantum field theory. Accordingly, physicists were thrilled to have finally taken such a big step towards precise findings and accuracy in predictions in the microcosm.

But in the end, mathematics did not offer as secure a foundation as the physicists would have liked. They repeatedly encountered difficult mathematical problems, some of which were perceived as insurmountable. For example, the calculations of interactions with virtual particles in the equations arising from the Feynman graphs often resulted in infinities that could

---

[4] Original: Richard Feynman, *Surely You're Joking, Mr. Feynman!: Adventures of a Curious Character*, W. W. Norton Publishing, New York (1985).

[5] N. David Mermin, *What's Wrong with this Pillow?* Physics Today, April 1989, page 9. This sentence is sometimes also attributed to Paul Dirac or Richard Feynman.

not be circumvented. Even the brilliant Dirac failed in his attempt to solve the problem of the infinities. Finally, a trick was developed that physicists call renormalisation. Just as Planck introduced quantised physical quantities in 1900 out of sheer desperation to somehow get ahead in his calculations, mathematicians introduced renormalisation to eliminate infinities. In fact, the physical quantities under consideration now assumed the desired finite values, and the calculated values even agreed with the experimentally measured values. Here, too, there is a parallel with Planck: he did not have "permission" for this intervention, but it worked.

Since there are no practically usable alternatives and their success proves them right, physicists and mathematicians still use this "dirty mathematics" to this day. However, they were not entirely comfortable with it, and the same can be said today. Most of them admit that renormalisation only sweeps the problem of infinities under the carpet. Paul Dirac remained a critic of renormalisation throughout his life. Even the pragmatist Feynman expressed restraint:

> The shell game that we play ... is technically called 'renormalization'. But no matter how clever the word, it is still what I would call a dippy process! Having to resort to such hocus-pocus has prevented us from proving that the theory of quantum electrodynamics is mathematically self-consistent. It's surprising that the theory still hasn't been proved self-consistent one way or the other by now; I suspect that renormalization is not mathematically legitimate.[6]

The computational methods of quantum field theory still do not exactly fit pure mathematics, so quantum field theory does not have a clean and contradiction-free mathematical foundation.

Particle Zoo Without Theory

With his anti-electron thesis, Dirac had predicted the positron; a few years later, experimental physicists were able to prove it. This sequence was reversed in the following decades. Unknown particles were found in experiments, and the mathematical models could barely keep up. The world consisted of much more than protons, neutrons, and electrons! All of a sudden, things started to get more complicated again.

---

[6] Feynman, Richard P., *QED, The Strange Theory of Light and Matter*, Princeton, Princeton University Press (1990), p. 128.

- During the measurement of cosmic rays, new particles appeared, including pions and muons.
- Particle acceleration, developed in the 1950s, is still a central activity of experimental physicists today: particles are brought to very high speeds and then made to collide with the help of electromagnetic fields. The higher the energies used, the more "fragments" are created and the deeper the physicists' insight into what we still call the "structure of matter".

Physicists began to discover more and more particles with ever more exotic properties and a new branch of physics was born: particle physics. The list of these exotic particles quickly grew to a dozen or more (today 61 different types of particles are known).

But why are there so many different particles in nature and what is their relationship to each other? While the theorists were working flat out to find an explanation, the number of experimentally detected particle types continued to grow. Every attempt to capture the characteristics of the particles already identified led to the discovery of new particles and thus to even more confusion. Soon the physicists ran out of names, so they attributed only ancient Greek (capital) letters to the new particles: Σ-, Λ-, Ξ-, and Ω-hyperons are just a few examples.

How could this multitude of particles be arranged into a system? Dmitri Mendeleev had solved a similar problem about a hundred years earlier when he classified the then known chemical elements at the time into eighteen groups of the periodic table, a system that is still valid today. Only one thing was clear to particle physicists: the classification of quantum particles would not be as descriptive as Mendeleev's.

In the mid-1960s, the American **Murray Gell-Mann**, another genius of twentieth century physics, succeeded in creating a first classification of the particle zoo. He approached the problem just as pragmatically as Mendeleev. First, he turned to the so-called hadrons, which are characterised by the fact that they are subject to the strong nuclear force, a basic physical force known since the 1930s. It is thanks to the strong nuclear force that the atomic nucleus holds together, despite the fact that the electromagnetic force tends to drive the protons apart.

The experimental measurements had shown that some hadrons hardly differ in their properties. For example, protons and neutrons have almost the same mass and spin. A neutron can even change into a proton, and a proton into a neutron, emitting a positron and an electron respectively. One of the few differences between protons and neutrons is their charge. Gell-Mann concluded that these two particles must belong to the same group.

He thus divided particles that differed more from protons and neutrons into another hadron group. Now there were two hadron types:

- Baryons, which include the proton and the neutron.
- Mesons, which include lighter particles such as the electron.

But could this division into baryons and mesons also be represented mathematically? Now Gell-Mann followed in the footsteps of the German mathematician Emmy Noether: he used the mathematical concept of symmetry groups, which describe the relationship between symmetries and conserved quantities (see the next chapter). One of the highly abstract groups found by Gell-Mann turned out to be perfectly suited for his purpose of sorting the observed particles. This was the group called SU(3), the "special unitary group of complex rotations in three-dimensional complex space". This is in fact an eight-dimensional group, and its representations could be used to classify the various hadrons into groups of eight. Some positions in these groups were still unoccupied, but over the years the gaps were filled precisely with newly discovered particles, and today, eight baryons and eight mesons are known. Gell-Mann's eightfold way created a first semblance of order in the particle jungle and today forms the foundation of elementary particle theory.

Particle physics is not the only area in which American research is still the undisputed leader in the world today—if you want to be at the forefront as a physicist, you go to the USA for at least a few years. Research in Göttingen and Berlin has never regained its former supremacy, and scientific life in Germany moved elsewhere after the Second World War: in Hamburg, the particle accelerator DESY came into being from 1959 onwards, and Bonn gained world-class importance in mathematics. Some European institutions conduct research at a similarly high level. Zurich, with the Swiss Federal Institute of Technology (ETH), has had an institution of world renown for over ninety years. The same applies to the English universities of Cambridge and Oxford, and the École Polytechnique and the Ecole Normale Supérieure in Paris deserve mention.

The **eight-dimensional SU(3) group** proved to be a mathematical structure using which hadrons could be classified into groups of eight. Its discoverer, the American Murray Gell-Mann, who had a wide range of interests, chose to call this classification the "eightfold way", a term that in Buddhism describes the noble path to the highest knowledge.

In the meantime, it is no longer a question of USA vs German-speaking countries, and nor is it just USA vs Europe. Other important players have won a place in the research community. Russia has produced very good theoretical physicists and mathematicians, some of whom have won Nobel Prizes. Japan is catching up fast, and India is strong in mathematics, but still lacks the resources for experiments of international importance. China has not yet been able to contribute any outstanding findings to basic research—the only Nobel Prize awarded to a Chinese citizen in a scientific subject was won by the physicist Tu Youyou in 2019. But the country is very good at translating theoretical findings into technological applications. For example, it has established a strong position in the field of artificial intelligence and the latest quantum technologies.

What will happen in the future? Perhaps it will once again be Europe that shows the world the way: CERN (*Conseil Européen pour la Recherche Nucléaire*) was founded near Geneva in 1957. It is located partly on Swiss and partly on French territory. And yet it is more than a purely European research centre: management, financing, and scientific recruitment take place at global level. This is because the deeper we want to penetrate into the basic structure of the universe, the greater the effort that has to be made. Today, such research rarely takes place in individual laboratories any more, but is managed by huge consortia that are internationally staffed and financed. The success of this approach to research, which no longer takes national borders into account, proves it right. One facility at CERN is the LHC (*Large Hadron Collider*), the world's largest and most important particle accelerator today. In 2012, the Higgs boson was detected here. Predicted in 1964, this is what gives particles their mass. For more than half a century, people all over the world had been trying to find this last missing building block of the Standard Model of particle physics, a model which explains all such phenomena to this day. A single country could not have made this effort. An attempt by the USA to build a much larger particle accelerator than the LHC was eventually dropped: in fact, in the 1990s, the construction of the planned Super-Conducting Supercollider (SSC) was cancelled for cost reasons.

9

Mathematics Becomes a Superpower—How Emmy Noether, John Von Neumann, and Alan Turing Changed the World

The fundamental transformation of physics began with Maxwell and Boltz-mann, whose thinking was still rooted in the nineteenth century. Einstein, Bohr, Heisenberg, Pauli, Dirac, Schrödinger, Feynman, and many others followed them and led the subject into a new era. They all accompanied their field through its deepest crisis, during which what had been believed to be certain and calculable turned out not to be so, showing that a new foundation would have to be found. The almost superhuman effort with which they succeeded in reinventing physics testifies to humankind's unstoppable quest for knowledge. Their work not only gave us a new world view that comes closer to reality than any other, but also made possible all the technologies that significantly shape our lives today.

Mathematics was going through similar crisis-ridden times. Famous mathematicians such as Cantor, Hilbert, and Gödel paved the way for new kinds of calculation of unprecedented abstraction. This process is even more significant because mathematics is the basis of physics and also of other sciences. At the same time, the level of abstraction developed in mathematics demanded more and more from (theoretical) physicists and, as we saw at the end of the last chapter, has ultimately left the mathematical description of quantum field theory unsatisfactory to this day. However, without the possibility of calculating and predicting phenomena with this increasing complexity, the technological leaps of the last decades would not have been possible, even beyond physics, de facto in any natural science.

© The Author(s), under exclusive license to Springer Nature
Switzerland AG 2022
L. Jaeger, *The Stumbling Progress of 20th Century Science*,
https://doi.org/10.1007/978-3-031-09618-1_9

But while physics can boast an outstanding hero in the public perception, in the person of Albert Einstein, such a leading figure is missing in mathematics. To those in the know, there is hardly any doubt about which mathematician was the "Einstein of his subject". Yet his name is known to only a few people today: John von Neumann. At his side were two other important mathematicians: Alan Turing and Emmy Noether.

The Unknown Universal Genius of the Twentieth Century

Every year, the *Financial Times* chooses a "Person of the Year". Bill Gates (1994), Barack Obama (2008), and Angela Merkel (2015) have already been honoured in this way; Mikhail Gorbachev even enjoyed this title twice, in 1985 and 1989. To coincide with the turn of the millennium in 1999, the editors of the *Financial Times came up with* something special: They chose a "Person of the Century". Politicians were deliberately excluded—Winston Churchill or Franklin D. Roosevelt, for example, would have been conceivable, for they had undoubtedly changed the course of the twentieth century. But the most profound events and advances had not occurred in the field of governance. It had been the century of natural sciences. So a scientist was honoured. The choice fell not on a celebrity like Albert Einstein or Werner Heisenberg, but on the Hungarian–American mathematician and polymath John von Neumann. Calling him the "Person of the Century" was more than justified. He had influenced mathematics like no other researcher since Newton. He gave it its present pure, abstract form and was at the same time a significant initiator of its concrete technological applications in all the natural sciences.

John von Neumann, born in Budapest in 1903, was the perfect example of a child prodigy. By the time he was six, he was dividing eight-digit numbers in his head and conversing in ancient Greek; he exchanged jokes with his father in that language. By the age of eight he was familiar with differential and integral calculus, and at twelve he was enthusiastically reading Emile Borel's *Théorie des Fonctions* ("Theory of Functions"), a classic of higher analysis[1]. By the time he was 19, he had already published two important mathematical works, the second of which contained the modern definition of ordinal numbers, which replaced the definition of the old master Georg Cantor and remains one of the foundations of number theory to this day. One of his

---

[1] E. Borel, *Leçons sur la théorie des fonctions*, Gauthier Villars, Paris (1898).

later teachers at the ETH in Zurich, George Polya, who was himself a great mathematician, once said:

> Johnny was the only student I was ever afraid of. If in the course of a lecture I stated an unsolved problem, the chances were he'd come to me as soon as the lecture was over, with the complete solution in a few scribbles on a slip of paper.[2]

> **Albert Einstein led the** way for several generations of brilliant scientists in the field of physics. Together, they fundamentally renewed their discipline, which had been shaped by the problems of the smallest particles and the mysterious ways of time.

> In mathematics, **John von Neumann** filled this leading role. He led the subject, which had been trapped in its own contradictions until the 1930s, out of the crisis and cleared the way for an age of mathematics.

In 1925, at the age of 22, von Neumann succeeded in solving one of the most pressing mathematical problems of the day in his doctoral thesis: he created a new axiom system for set theory, which suffered from Russel's antinomy, avoiding the contradictions of his predecessors. Only later did mathematicians realise that he had thus also given the *whole of mathematics* a new and reliable foundation (this Chapter). Gradually, he made a name for himself as an outstanding mathematician. When Werner Heisenberg gave a lecture in Göttingen in 1926 on the difference between his and Schrödinger's theories of quantum mechanics, the old master Hilbert asked his physics assistant Lothar Nordheim what on earth this young man had been talking about. Nordheim sent his professor a paper with a summary that Hilbert still did not understand. What happened next was described by Nordheim thus:

> When von Neumann saw this, he cast it [Heisenberg's theory] into an elegant axiomatic form in a few days, much to Hilbert's liking.

---

[2] Miodrag Petković, *Famous puzzles of great mathematicians*. American Mathematical Society (2009) p. 157; https://archive.org/details/famouspuzzlesgre00mpet/page/n175/mode/2up.

Von Neumann was only 23 years old at the time and had only been in Göttingen for a few weeks. As a newcomer, he had immediately demonstrated his talent and, to Hilbert's delight, made extensive use of the concept of the so-called Hilbert spaces.

How did von Neumann achieve these fantastic feats at such a young age? His success was certainly due to his incredible ability to break down complex mathematical issues into simple questions at lightning speed and then find answers to them off the cuff—what he couldn't solve quickly, he usually couldn't solve at all. In addition, he had a photographic memory and could recall complete novels and telephone directory pages on command. Over the years, he accumulated an encyclopaedic knowledge of all kinds of subjects. When von Neumann was in his thirties, a Princeton professor and expert on Byzantine history declared that the mathematician had greater expertise in this field of history than he did. And the physicist and Nobel Prize winner Hans Bethe once said about him:

> I have sometimes wondered if a brain like von Neumann's doesn't indicate a species superior to humans.[3]

Even Einstein, with whom von Neumann worked for over twenty years at the *Institute of Advanced Studies* in Princeton, did not possess von Neumann's analytical and mathematical intelligence. Einstein's insights were more profound than von Neumann's, but this was probably the reason why he had to spend ten years thinking about the general theory of relativity and almost failed because of its abstract mathematics.

A Celebrated Bon Vivant

Unlike some of his scientific colleagues, who were eccentric loners or mentally unstable, von Neumann was an extrovert and the centre of every society. He knew neither self-doubt nor competitiveness, and he was always so far ahead of others that he did not have to fear disparagement—he was simply almost always right. In heated discussions with other scientists, he was always concerned with the matter at hand. When Gödel presented his mathematical proof to a small circle in Königsberg in 1930, showing that the validity of axiom systems can never be proven, von Neumann sat at the table with him and immediately recognised the significance of this proof: even for his

---

[3] C. Blair, *Passing of a Great Mind*, Life: 89–104 (25 February 1957); https://books.google.ch/books?id=rEEEAAAAMBAJ&pg=PA89&redir_esc=y#v=onepage&q&f=false.

own axiom system, which he had presented to the scientific community five years earlier, he could now no longer hope to prove it one day. Instead of resisting Gödel's discovery, he immediately accepted the mathematical truth. What was more, von Neumann immediately recognised that Gödel's incompleteness theorem could be extended, and only a few weeks later, he had formulated the extension. However, he refrained from submitting his work, and instead brought it to Gödel. Gödel politely thanked him and informed him that he had already submitted precisely this addition for publication. If von Neumann had attached importance to this, one of the most important theorems of modern logic would probably not be called "Gödel's Second Incompleteness Theorem" but "the Von Neumann–Gödel Theorem".

In the following years, von Neumann continued to work on the foundations of mathematics, i.e., precisely at the core of the crisis into which the field had slid: the mathematical assumptions on which classical physics had been based for centuries had turned out to be unsustainable. In 1932, in his book "Mathematical Foundations of Quantum Mechanics", he created the first consistent mathematical foundation of the new quantum theory. Before that, quantum theory had contained equations whose solutions agreed with the results of experiments, but there was no solid mathematical formulation. For example, Born and Schrödinger had already recognised that Schrödinger's wave function and Heisenberg's matrices were equivalent, but their reasoning did not satisfy the demands of mathematical rigour. Only von Neumann's new foundations were able to prove the equivalence of the two approaches in a mathematically exact way.

> The **mathematical description of quantum theory** is still based on the concept of linear operators in infinite-dimensional Hilbert spaces formulated by John von Neumann and David Hilbert.

Von Neumann also made outstanding contributions in other areas of pure mathematics. Among these:

- He formulated the foundations of so-called **quantum logic**, creating a whole new field of research that attempted to reconcile the apparent contradictions of classical logic (for example, the liar paradox from antiquity) with the impossibility of simultaneously measuring position and momentum.
- He made fundamental contributions to the **ergodic hypothesis**, a branch of mathematics that deals with the states of dynamical systems. His

statistical ergodic theorem is still one of the foundations of statistical physics.

Von Neumann's contributions to *pure mathematics* alone would probably be enough to make him one of the most important mathematicians of the twentieth century. But he achieved his greatest impact in the *concrete application* of mathematics. Thus, he influenced the entire landscape of the natural sciences in the second half of the twentieth century. Here are a few examples:

- Von Neumann considerably raised the intellectual and mathematical level of **economics**. Among other things, he founded game theory, which is now an important field of theoretical economics.
- He developed mathematical methods that are still of great importance in **engineering** today. When George Dantzig, an expert in linear processes, presented von Neumann with an unsolved problem, he simply said, "Oh, that!" and gave an impromptu lecture lasting more than an hour, explaining how the problem could be solved with the help of the "theory of duality" he had spontaneously developed on the spot.
- He developed a mathematical analysis of **biological organisation and self-reproduction**, as well as the evolution of complexity, which is fundamental to **biology**. His "theory of cellular automata" is based solely on mathematical simulations and made a significant contribution to the description of gene replication, preceding the discovery of the structure of DNA by James Watson and Francis Crick by several years. Von Neumann himself once said:

It is clear that the copying mechanism performs the fundamental act of reproduction, the duplication of genetic material, which is clearly the fundamental operation in the reproduction of living cells.[4]

In 1887, Boltzmann formulated the statistical rules according to which particles move in space in the **ergodic hypothesis**. These considerations gave rise to what is known as chaos theory in the 1980s, among other things.

---

[4] J. von Neumann, *The General and Logical Theory of Automata*; in L. A. Jeffress (ed.), *Cerebral mechanisms in behaviour; the Hixon Symposium* (p. 1–41), Wiley, London (1951).

A few sentences later, he anticipated a research direction that is only now getting under way: the self-reproduction of nano-machines, which will in all likelihood bring about major changes in medicine in the near future:

> Small variations of the above scheme also allow us to construct automata that can reproduce themselves and, moreover, construct others.

John von Neumann was thus a pure mathematician whose achievements remain unrivalled to this day. At the same time, his applied mathematics helped researchers in almost every field to make breakthroughs, from economists to electrical engineers. In the life of von Neumann there was even room for a third function, in which he *quite directly* reset the course of world history.

The Architect of the Red Button

In the late 1930s, von Neumann became interested in explosions. These are difficult to model due to the speed of the processes involved and the turbulence that occurs. He quickly became the leading authority on the mathematics of shock waves. This knowledge made him interesting to the military and he soon held several military advisory roles in parallel. With the entry of the USA into the Second World War in 1941, his activities took on special significance. From 1943, as mathematical director of the Manhattan Project, he took a central role in the construction of the American atomic bomb:

- It was von Neumann's calculations that showed that, not only was a certain uranium isotope fissile, but that enriched plutonium could also be used in atomic bombs.
- He created a concept for the calculation of explosive lenses and was responsible for their actual installation in plutonium bombs. Without his skills, the USA would have had only one uranium atomic bomb at its disposal in the summer of 1945 and, at that time, the atomic bomb project would not have progressed beyond a first test in the New Mexico desert.
- He supervised the calculation of the forces unleashed and calculated the height above the ground at which the bombs had to be detonated for maximum effect. This activity made him a member of the *Target Committee*, which deliberated on the targets of the atomic bombs in Japan. If von Neumann had had his way, the first atomic bomb would have been

dropped on Berlin. He was extremely annoyed that Germany had surrendered before the atomic bomb was ready for use. Now the only enemy left was Japan. The first atomic bomb fell not on the old imperial city of Kyoto, as von Neumann had suggested, but on Hiroshima. The target of the second bomb was originally planned to be the city of Kokura, which had developed into a centre of the arms industry since the mid-1930s. Since it was too cloudy there that day, the bomb was dropped on nearby Nagasaki.

- After the Second World War, von Neumann was instrumental in the further development of the American nuclear weapons programme, which developed the hydrogen bomb (H-bomb), and in the US missile programme for the development of intercontinental nuclear explosive devices.

- To model nuclear weapons, especially the H-bomb, he developed methods for solving what are known as hyperbolic partial differential equations. He had already come across this type of equation during his work on fluid dynamics and hydrodynamic turbulence. Since these equations could not be solved by classical methods, he developed the **Monte Carlo method**, which is widely used today. In hundreds of runs—today there are many thousands—the process is calculated with different values for the parameters. These simulations produce a probability distribution that is very close to the exact result. The first application of this method was the calculation of neutron diffusion in the explosion of an atomic bomb.

How does a person cope with being the direct author of bombs that cost hundreds of thousands of lives and brought much suffering to an entire country? Von Neumann saw no problem in destroying enemies. For him, there were "the good guys" who were clearly distinguishable from "the bad guys". In military circles, he was revered as a hero because he had saved the lives of many Americans by forcing Japan to surrender soon after the bombings. In public, however, his activities as one of the leaders of the Manhattan Project were hardly known.

An atomic bomb is detonated by converting a previously uncritical mass of fissile material into a critical density state. This can be done in two ways:
1. Two uncritical masses are shot at each other inside the atomic bomb, so that a chain reaction is set in motion. A suitable material is the naturally occurring but very rare uranium isotope U^{235}, which has to be supplied in large quantities. The Hiroshima bomb was of this type.

2. In the implosion principle, **explosive lenses** are installed around a sphere of matter. When ignited at the same time, the material is compressed and a chain reaction begins inside the sphere. In this type, plutonium can be used in addition to uranium U^{239}. The production of this isotope, which does not occur in nature, is not quite as complicated as the enrichment of a sufficient quantity of U^{235}, but the construction of a plutonium bomb is much more difficult. The test bomb Trinity and the bomb dropped on Nagasaki were plutonium bombs.

The Computer Sees the Light of day

During his work on the hydrogen bomb, von Neumann encountered a problem. The calculations were becoming too time-consuming! Richard Feynman, who had worked with him on the development of the atomic bomb at Los Alamos, described the procedure at the time like this:

> That's why we occupied a room with women. Each had a Marchant [a mechanical calculating machine]; one multiplied, the next added. Another cubed – she did nothing but cube the number on an index card and then pass it on to the next girl.[5]

These mathematical assistants were called "computers". However, it was no longer possible to perform the escalating number of calculation steps by hand. So von Neumann looked into the possibility of having this work done by electronic machines. Some electronic machines already existed. But here, too, the procedure was tedious and cumbersome: the programme was permanently installed as hardware in the computer. Depending on the desired calculation steps, drawer-sized plug-in modules had to be installed in the main unit. This meant that the computer had to be shut down for every programme change. This only made sense after all the computing steps had been completed, because there was still no way to store data. If the scientists realised that they had forgotten a step in the calculation, they had to start all over again, and reprogramming could take several days.

The **Von Neumann architecture** was the decisive breakthrough in the development of programmable calculating machines. In computers built according to

---

[5] Richard P. Feynman "You must be joking. Mr. Feynman!" Piper Munich, 6th edition (1991).

this principle, data and program are binary coded in the same memory. This means that the program can also be changed while a process is running.

In order to make complicated differential equations accessible to machine solution, von Neumann developed a new computer architecture, thus becoming a pioneer of computer technology in addition to all his other outstanding achievements. In 1945, von Neumann published the *First Draft of a Report on the EDVAC*—the acronym stands for *Electronic Discrete Variable Automatic Computer*—in which he described the von Neumann architecture that is now named after him. From 1949, he led his own computer project at the *Institute for Advanced Study* in Princeton, where Einstein also worked, and developed the IAS computer, the first universal computer, i.e., a computer that can process any computational program with one and the same hardware. His programming concepts and the von Neumann architecture became the basis for all modern computers. Specifically, the architecture consists of the following building blocks:

- A **processing unit** consisting of a logic unit that processes commands such as AND, OR, IF/THEN, and a processor register that serves as a temporary buffer for solutions of partial steps.
- A **control unit** consisting of a command register (program) and a program counter that ensures the correct sequence of program steps.
- A **memory unit** that stores data and commands.
- An **external memory**.
- Input and output mechanisms for the programme and the data to be processed.

Even though today the machine solution of complex differential equations is indispensable for weather forecasts and climate modelling, among other things, it should not be forgotten that the original purpose of von Neumann's computer and its algorithms was to perform calculations on the internal processes of hydrogen bomb explosions.

Alan Turing—The Birth of the Digital Age

Einstein had Niels Bohr as a colleague and friend with whom he could discuss the new quantum theories. For von Neumann, Alan Turing was the sparring partner; he was one of the few people who could follow him on his intellectual flights of fancy, and he was the second important pioneer of the twentieth century in the field of applied mathematics. An eccentric loner, Turing was a very different type from the urbane von Neumann—for example, he chained his coffee mug to the radiator to make sure it would not be stolen and rode a bicycle wearing a gas mask because of his hay fever. A talented long-distance runner, he achieved world-class marathon times. When he was working on decoding the German Enigma machine at Bletchley Park during the Second World War, he occasionally ran the 64 km to London to attend meetings in the capital. In 1948, he had a good chance of making the British Olympic team, but an injury prevented him from qualifying.

Together with von Neumann, he stands at the beginning of modern information technology. He only actually wanted to solve the "decision problem", one of Hilbert's 23 open problems (see page 49):

- This asks whether, for *every statement* in mathematics, there is a way to classify it as true or false.
- The alternative is that there might be, for example, arithmetical problems for which you have to keep calculating "forever" without ever knowing whether there is actually a solution (and whether it is worth the effort to keep calculating).

In order to be able to finally solve the decision problem, Turing developed a machine in a thought experiment that can perform any calculation that can be handled by an algorithm with only a few basic operations. Using this idea, now known as the Turing machine, he showed that there is a so-called halting problem with certain calculations. For example, if the machine has to compute the number Pi, it will never stop because Pi is a non-periodic, infinite decimal number. This illustrates the connection between the stopping problem and the decision problem:

- No one knows at the beginning of the calculation whether the machine will come to an end at some point in the processing of a particular task or whether it will continue to run forever.
- This makes it clear that some statements in mathematics cannot be definitively classified as "true" or "false". Turing had thus found a very simple and at the same time elegant answer to the decision problem.

Turing machine: In 1936, Alan Turing described a computing machine whose elements only know two states: "0" and "1". The fact that even the most difficult arithmetic operations can be performed on this basis is one of the foundations of theoretical computer science today. The more complicated the calculations are, the more logical links there are and the more zeros and ones have to be processed one after the other. A few years after Turing's thought experiment, a universal computer working according to this scheme and able to calculate all mathematical problems was actually built.

Turing published his findings in 1936 under the title *On Computable Numbers, with an Application to the Decision Problem.*[6] Von Neumann's central concept of the modern computer was based on this article, something he openly acknowledged. Turing's work became compulsory reading for von Neumann's co-workers in the EDVAC project. In short, one could say:

- Turing invented the Turing machine,
- von Neumann built it.[7]

Another passion shared by von Neumann and Turing was the application of mathematical principles to biology. While von Neumann worked on self-replicating biological systems, Turing was interested in modelling reaction–diffusion processes that describe the emergence of biological patterns in living beings. The fur markings of big cats and zebras, for example, can be seen quite figuratively as a biological pattern. It is obvious that *there is* a pattern, but which mechanisms determine exactly which stripe arrangement emerges? There are countless ways to "paint" stripes on a zebra. Turing believed that comparatively simple but non-linear laws produce the diversity and complexity of life. There were also complicated differential equations in this field that could only be solved by using computers.

Alan Turing did not become the "Person of the 20th Century", but he was included in *Time Magazine*'s 1999 list of the 100 most important people of the twentieth century. The reason given was[8]:

The fact remains that everyone who taps at a keyboard, opening a spreadsheet or a word-processing program, is working on an incarnation of a Turing machine.

---

[6] A. Turing, *On Computable Numbers, with an Application to the Decision Problem*, Proceedings of the London Mathematical Society (1937).

[7] For a detailed account of the historical development of the evolution of the first computers, see: G. Dyson, *Turing's Cathedral: The Origins of the Digital Universe*, Vintage, New York (2012).

[8] TIME Magazine, Vol. 153 No. 12 (29 March 1999); Time 100 People of the 20th Century.

Turing Test: Turing was one of the first computer scientists to address the question of artificial intelligence. He suggested an experiment that is known today as the Turing Test: A computer is said to be capable of thinking or intelligent if a human interlocutor cannot recognise that he or she is communicating with a machine.

For more than two decades, Turing and von Neumann met irregularly. Without the exchange between them, the disciplines of mathematics and computer science, and with them the entire computer age today, would probably not be where they are today. However, one can also view the matter from another perspective: Turing committed suicide in 1954 at the age of 41, von Neumann died of cancer three years later at the age of 53. During their scientific careers, they had never stopped working on the idea of ever more powerful computing machines. As early as the 1930s, at that time Alan Turing spent almost two years at Princeton, where von Neumann also worked, they jointly developed a philosophy of artificial intelligence. In 1950, Turing formulated the Turing test named after him, and in the following years von Neumann researched using analogue codes that allow *parallel* treatment instead of binary 0/1 codes that have to be processed step by step *one after the other*. It is precisely this approach that has been playing a role again more recently in the development of artificial intelligence and neural networks. Who knows? If Turing and von Neumann had lived longer, our computers today would be much more powerful.

The Mathematicians' War

Just like von Neumann's skills, Turing's were also in demand during the Second World War. The First World War had been a war of chemists. The Second, while it was still going on, was called the "war of the physicists"—the atomic bomb and radar are key words here. And it was Albert Einstein who wrote a letter to the American President Franklin D. Roosevelt on the necessity of building an atomic bomb. His concern that the physicists who had remained in Germany might succeed in building such a weapon drove him to take this step, despite all his misgivings. It is less well known that mathematics played a role in deciding who would win and who would lose, and later contributed to the global supremacy of the major blocs. Von Neumann and Alan Turing, the two most important and brilliant representatives of their field, played a decisive role in ensuring that their two countries, the USA and England, emerged victorious from the Second World War.

The Briton Turing worked as a mathematician at the *Government Code and Cypher School* in Bletchley Park during the Second World War, where he succeeded in deciphering German codes. Among other things, he developed the electromechanical machine that tracked down how the German *Enigma* coding machine worked. By successfully deciphering encrypted messages from the German military, he was instrumental in the Allied victory. Some estimate that, by deciphering the Enigma code, he shortened the war in Europe by more than two years and saved over 14 million lives.

Soldiers of science. A country's performance in the research fields of mathematics and physics had an all-important military significance during the Second World War. It retained this role in the ensuing Cold War.

Like von Neumann, Turing continued to work on the design of electronic computers after the war.

- In 1946, he presented the first concrete design for a machine based on the von Neumann architecture. Von Neumann had published a similar, less detailed work some time earlier, building on "a number of ideas that came from Dr. Turing" (according to John R. Womersley, then head of the mathematics department at the National Physical Laboratory, NPL, in London).
- In September 1948, Turing developed a coding scheme on a 5-bit basis for the *Manchester Mark 1*, one of the first computers that could store its programming. The 32 possible configurations of the 5 bits (from 00000 to 11111) allow a direct assignment of the 26 letters. With this standardised code, programmes and data could be recorded on punch cards, transferred, and then read out again. This system was used until the 1970s. After that, code with a 6-bit basis was used, allowing 64 characters. Today, an 8-bit basis is mainly used, allowing 256 characters.

The achievement of the two mathematicians was widely recognised: von Neumann received the *Presidential Medal of Freedom*, the highest award for civilians in the USA, and Turing the *Order of the British Empire Award*. The latter did not prevent British society from socially ostracising its celebrated war hero Alan Turing only a few years later. Due to his homosexuality, which was still considered a crime in the mid-twentieth century, Turing was sentenced to one year in prison by the British judiciary for gross immorality in March 1952. As an alternative, he was offered hormone treatment, including synthetic oestrogen. The physical changes that set in, combined by the way he was hounded by the authorities, were probably what caused him to fall into a depression. Two years later, on 7 June 1954, he committed suicide with cyanide.

Since 1966, the *Turing Award* has been considered the highest distinction to those who have achieved great things in computer science—it is, so to speak, the Nobel Prize for computer scientists. Nevertheless, his official rehabilitation had to wait until 2009.

Emmy Noether Finds a Compass for the Abstract World

The great twentieth century mathematician John von Neumann is known today almost only to insiders, even though his genius was briefly brought to the attention of a wider public by his election as a "Millennium Person." Turing was also all but forgotten until the tragedy of his life became widely known through the 2014 film *The Imitation Game*. Today, almost everyone in the world benefits on a daily basis from the incredible achievements of these two mathematicians, and yet their memory is far from being adequately honored.

Symmetry: The human body serves as a model for harmonic relationships. Leonardo da Vinci's famous depiction illustrates this explicitly. With outstretched arms and legs, a square and a circle can be drawn around the human figure, centered on the navel.

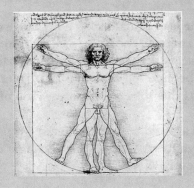

But there was a third mathematician—this time a woman—for whom even these late honours have been denied. Although she enriched theoretical physics with a theorem that sets what is arguably the most beautiful and sublime principle in the natural sciences into mathematical form, the German mathematician Emmy Noether remains unknown outside the professional world. Noether's theorem, named after her, establishes the connection between **symmetries** in mathematics and **conserved quantities** in nature. Both terms will be explained in the following.

The word **symmetry** comes from the ancient Greek *symmetría* and means evenness or uniformity. In almost all views of art throughout human history, it has been considered an essential criterion for beauty and perfection. Sculpture, painting, and architecture strive for its ideal proportions. Symmetries also play a central role in modern physics. Here it is not a matter of relative proportions, equilibria, or reflections, but rather of natural order and structure. Physicists cherish the deep belief that the laws of nature will prove to be relatively simple and manageable in principle, despite the diversity of the phenomena they must govern. The beauty of nature is revealed in this simplicity. This is what Werner Heisenberg says about symmetry in modern physics:

> The final theory of matter will be characterised, similarly to Plato, by a series of important symmetry requirements. [...] These symmetries can no longer be explained simply by figures and pictures, as was possible with Platonic solids, but they can be explained by equations.[9]

Symmetry gives rise to the two most important conservation laws of physics, which impose stability on ourselves and our environment:

- It is only because the laws of nature do not change over *time* that the law of conservation of **energy** applies: in a closed physical system, the total energy never increases or decreases, but always remains the same.
- Likewise, it is only because the equations of physics do not change under *spatial* displacements that the law of conservation of **momentum** applies.

We cannot imagine a world without symmetry—pure chaos would reign. This is because only symmetry ensures that there are reliable laws of nature that apply always and everywhere:

---

[9] W. Heisenberg, *Across the Frontiers*, Harper & Row, London (1974).

- A stone falls to the ground today according to the same rules as it did yesterday and will do tomorrow.
- Likewise for displacements in space: the same physical laws apply everywhere in the world.

Symmetries in mathematics are even more abstract than in physics. Equations that retain their shape under certain changes in their variables (just as the laws of nature do, even when time and place change) are said to be "symmetric" by mathematicians. In the nineteenth century, the Frenchman Évariste Galois and the Norwegian Sophus Lie developed a new mathematical discipline called group theory, the former for algebraic equations, the latter for differential equations and general geometric structures. A mathematical group is a set of elements with certain properties. In the case of **symmetry groups**, we consider the set of possible variable changes (transformations) that are possible without changing an object or an equation. In mathematical terms, a symmetry group includes all transformations that leave geometric entities, equations, or other mathematical objects invariant. In addition, there are certain linking rules that define the interconnection of such transformations and are themselves part of the group.

A simple example of a **symmetry group in mathematics**: A two-dimensional, regular polygon with n sides is rotated in the plane. Every 360/n degrees, we obtain an image that looks exactly like the starting position. If, for example, a hexagon is rotated by 60°, it looks exactly as it did at the beginning. For all rotations that are a multiple of 60°, the same thing happens. These rotations form a group. Any elements from this group can be linked together—for example 60° plus 120°, or 240° minus 120°—and the hexagon does not change its appearance in space.

This is where Emmy Noether comes into play. She showed that the mathematical symmetries in physics are connected to **conserved quantities.** These are physical variables that do not change their value when the system is subjected to certain changes. She proved that a certain symmetry can be assigned to *each* physically conserved quantity, and vice versa. For physicists, this fundamental connection provides a great advantage: starting from a known mathematical symmetry group, they can specifically set out to find new conserved quantities that will lead them to new laws. The reverse is also possible: the underlying symmetry can be deduced from known conserved quantities, and this in turn paves the way for the discovery of further conserved quantities.

Proving the connection between symmetries and conserved quantities had not been much more than a little finger exercise for Noether, because the associated mathematics is not very complex in its original version. Einstein found the new theorem very interesting and praised Noether for her discovery, but she herself would hardly have called it one of her outstanding achievements. In the following years she showed far greater mathematical genius. Among other things, she gave algebraic geometry its supporting foundation with new axioms, similarly to what von Neumann had done for set theory. Emmy Noether should actually be known and honoured as the "mother of modern algebra", but this recognition has been denied her right up until today. Perhaps this is because the structures she established are so abstract that only a few mathematicians can grasp them. Among theoretical physicists, on the other hand, she has been revered since the 1960s because of the mathematically relatively simple Noether theorem.

If physicists had to name the most decisive principle of the natural sciences, the majority would probably name the "Noether theorem", which pairs symmetries with conserved quantities and is the basis of the Standard Model of particle physics today.

Symmetries: It is not only solid bodies that can be symmetrical. Abstract mathematical equations can be, too. Equations whose values do not change under certain changes in the variables are said to be symmetrical. In physics, symmetries—which can reach a very high degree of complexity—indicate a fundamental, structural order.

Conserved quantities: Examples are laws of nature that remain the same in the case of spatial or temporal shifts. In particle physics, the conserved quantities are defined in a much more complex way.

One can see it as a just reward that Emmy Noether has nevertheless gone down in the history of the natural sciences as one of the most important mathematicians. For in the second half of the twentieth century, many years after Noether's death in 1935, her incidental theorem set off on an unexpected career: it proved to be an indispensable tool for the increasingly abstract and complex mathematical physics of quantum field theory:

- The symmetry transformations in **classical physics** include the simple temporal or spatial shifts mentioned above.
- In the abstract spaces of **quantum field theory**, for example, the existence of the eight gluons of the strong nuclear force results from the connection between conserved quantities and symmetries.

It was only thanks to Noether's theorems that Murray Gell-Mann had been able to find the eight-dimensional SU(3) group that represents the symmetry of the strong nuclear force, and on the basis of which he was able to arrange the quantum particles into a system and develop the first features of the Standard Model of particle physics. Gell-Mann himself named the theorem "Noether's theorem" in her honour. The theoretical physicist Frank Wilczek from MIT (Nobel Prize in Physics 2004) said this about Noether's theorem:

This theorem has been a guiding star to 20th and 21st century physics.

Cinderella at the Mathematical Institute

Who was Emmy Noether? She was born in 1882 as the eldest daughter of the important mathematician Max Noether. Her abilities did not stand out at school, and it would have been very unusual for anyone to look for mathematical talent in a girl. Noether's career aspirations corresponded to the conventions of the time: she wanted to become a language teacher. In 1900, she passed the state examination for English and French. But instead of going to one of the Bavarian girls' schools as a teacher, she caught up on her Abitur and studied mathematics at the University of Erlangen, where her father was a professor. In 1907 she received her doctorate *summa cum laude*. The path to a university career was open to male fellow students with this top grade, but Emmy Noether was denied an academic position as a woman. She worked for her father at the Mathematical Institute for seven years without pay, including supervising doctoral theses that were officially supervised by her father. She was not "Emmy Noether", but "the daughter of Max Noether".

Against all odds, Noether's skills prevailed; her reputation grew with her publications.

- In 1908 she was elected to the *Circolo Matematico di Palermo*, at that time the most important association of mathematicians in Italy.
- In 1909 she was invited to become a member of the German Mathematics Society.

- In the same year, she was allowed to give a lecture at the society's annual conference in Salzburg. Numerous other invitations followed.
- In 1915, the greatest mathematician of the time, David Hilbert, and his colleague Felix Klein, took notice of her. The two persuaded Noether to join the mathematics department at the University of Göttingen, an international centre for mathematical research at the time.

Although the great David Hilbert advocated Noether's habilitation in Göttingen, the philosophy faculty, which was jointly responsible, refused to grant the special permission needed for women. Hilbert could not get his way, and even his famous remark "Gentlemen, this is not a bathhouse!" could not change the negative attitude of the established professors. (He was referring to the obligatory separation of the sexes in swimming pools at the time; whether Hilbert really said this is another matter, although it seems quite likely).

Einstein's competitor: In 1905, Einstein had published his *Special Theory of Relativity*. Ten years later, David Hilbert in Göttingen and Einstein in Berlin worked in parallel on its further development into the *General Theory of Relativity*. Hilbert submitted his work to the Göttingen Royal Society of Sciences for publication five days before Einstein. However, the proofs of this work that have survived today do not contain the decisive field equations; other documents, which unfortunately have not survived, are said to have contained them. So was Hilbert actually the originator of general relativity theory? No, because Hilbert's publication only covered an extremely complex mathematical aspect which he had solved with the help of tensor analysis, something he mastered better than Einstein. Hilbert had presented part of the theory of relativity in a very abstract way, but had not recognised the explosive power of his calculations. Einstein, on the other hand, intuitively grasped the significance of a four-dimensional space–time structure for physics.

Noether's career stalled again; for four years she had to hold her lectures under Hilbert's name. It was not until after the First World War that women were officially allowed to hold the rank of professor. Noether habilitated in 1919. Despite the new laws, women continued to have a hard time in academia, as shown, among other things, by the fact that Noether did not receive a salary even as a private lecturer. Although she was already an internationally recognised mathematician at this point, it took several more years before she was finally able to step out of her father's shadow. In the end, her genius prevailed: when Max Noether died in 1921, he was described as the "father of Emmy Noether".

It was Noether's patrons David Hilbert and Felix Klein who brought her to the topic of symmetries. The two mathematicians were studying Einstein's general theory of relativity (GR) and had come across a puzzle. When they tried to set up an equation for the conservation of energy, it led to an apparent tautology. It was like writing: "$0 = 0$". Such an equation has no deeper meaning. Hilbert and Klein were surprised, because for all other areas of physics, energy conservation laws could be formulated and used for calculations. Why was this not possible within the framework of relativity theory?

Hilbert and Klein turned to the then 36-year-old Noether, because she had the necessary expertise in the relevant areas of mathematics. In fact, she was able to show very quickly that this seemingly strange behaviour of the conservation equation in GR is typical for all covariant equations.

Emmy Noether was not only an outstanding mathematician. She is also a cautionary example of the fact that no society can afford to arbitrarily deny the abilities of a certain part of that society.[10] It is above all thanks to Noether's assertiveness that she was able to make her way for decades in a world dominated by men, despite all the opposition. Less self-confident women would have given up or not even dared to try to gain a foothold as a mathematician in the very traditionally minded university landscape. Just imagine, when Emmy Noether published her theorem in the Göttinger Mathematische Nachrichten in 1918,[11] she had long since demonstrated her extraordinary ability. Nevertheless, she was denied the opportunity to present her work in person to the Göttingen Mathematical Society. This had to be done by her mentor Felix Klein, because as a woman, Emmy Noether was excluded from this society.

> **Covariance** means that under certain changes of the variables the form of an equation does not change (for those interested in mathematics, covariance is expressed in an infinite-dimensional Lie algebra of the associated group).

> Noether's discovery that all covariant equations are invariant, i.e., have this particular symmetry, is known today as **Noether's second theorem**. To prove this, she formulated another theorem shortly afterwards that describes the connection between symmetries and conserved quantities. This is the theorem

---

[10] See Lars Jaeger, *Emmy Noether. Ihr steiniger Weg an die Weltspitze der Mathematik*, Südverlag (2022).
[11] E. Noether, *Invariant variation problems*, Gött. Nachr, pp. 235–257 (1918).

that is of such great importance for physics, and is therefore called **Noether's first theorem** today.

- In classical Newtonian physics and also in quantum theory, the equations are not generally covariant. If a force acts on a ball, then it is accelerated and it changes positions. If a ball is accelerated strongly, for example, the position variables are accelerated strongly, which leads to corresponding changes of its movement.
- In GR, the equations are covariant, so their results behave strangely invariantly and give us results we are not used to. This is because in this system there are no gravitational forces acting directly on bodies; it is space–time itself that changes (and therefore the movement of the bodies) due to gravitation produced by the presence of masses.
- The equations of the highly speculative string theory also exhibit this property.

10

The Architecture of Life is Decoded—How the "Science Clowns" Watson and Crick Ended a Decades-Long Quest

The insights of mathematicians took physics and the technologies that emerged from it to great heights. The new possibilities for calculation also radiated to other fields of science. For example, biology was increasingly perceived as an area that could be better understood using the same mathematical methods as physics. As a result, more than a few physicists and mathematicians began to work on the science of life in the 1930s.

- The quantum physicist Erwin Schrödinger was fascinated by questions about the nature of heredity, information storage in biological systems, and a possible genetic code. In 1943, he gave a lecture in Dublin entitled "What is life?", which became a book with the same title a year later.
- The mathematician John von Neumann analysed the algorithms of self-replication in cellular automata, which he used to simulate the reproduction of living systems in the first computer models. With their help, he also dealt with the increasing complexity of the then still unknown structures responsible for heredity through spontaneous mutations.
- Alan Turing was interested in random pattern formation in living beings. According to what rules, for example, do cells organise themselves into orderly associations? And according to what rules does evolution take place?
- Born in Berlin, Max Delbrück, a great-grandson of the chemist Justus von Liebig, began his scientific career as a theoretical physicist. He was a student of Pauli and Bohr and worked with the nuclear physicists Lise

© The Author(s), under exclusive license to Springer Nature
Switzerland AG 2022
L. Jaeger, *The Stumbling Progress of 20th Century Science*,
https://doi.org/10.1007/978-3-031-09618-1_10

Meitner and Otto Hahn from 1932. Niels Bohr suggested that he turn to biology. In 1935, Delbrück, together with the Soviet geneticist Nikolai Timofeyev-Ressovsky and the German physicist Karl Günther Zimmer, defined genes as complex atomic assemblies. You can read more about him on one of the following pages.

- Richard Feynman spent his sabbatical in 1959/1960 doing research on mutant viruses to learn more about the pathways of trait expression.

It is striking that the researchers were so strongly attracted to the topic of heredity. The mechanism whereby individual characteristics find their way from one generation to the next seemed to them to be one of the last great mysteries of the natural sciences. A major obstacle to solving this mystery was the lack of a mathematically consistent probability calculation that could have been used to explain genetic and hence evolutionary processes:

- Although Darwin based his entire theory of evolution on random mutations in 1859, he was far from backing up his statements with mathematical methods. This was probably also due to the fact that mathematics was not one of his strengths. In his autobiography, he wrote one year before his death:

My power to follow a long and purely abstract train of thought is very limited; and therefore I could never have succeeded with metaphysics or mathematics.[1]

- Mendel, who published his crossbreeding experiments in 1865, was, as a trained physicist, clearly more mathematically adept than Darwin, but he too lacked the mathematical tools to calculate probabilities exactly.

In the late nineteenth century, the physicists Boltzmann and Maxwell lacked the decisive argument that observable phenomena could actually be explained by random processes at the atomic level. Boltzmann had already come a long way with his concept of entropy and the associated quantification through the distribution of random configurations, but decisive steps were still missing. Hilbert, who knew him and his work well, overcame his aversion to chance and, at the Second Mathematical Congress in Paris in 1900, placed the lack of a systematic description of probabilities sixth on his

---

[1] Ch. Darwin, The Autobiography of Charles Darwin: 1809–1882, W W NORTON & CO; Revised Edition (1993).

list of 23 open mathematical problems. He had come to the conclusion that a complete explanation of the world by physics and thus the further progress of science would not be possible without the mathematical mastery of chance.

Randomness: Whether a single random event occurs or not cannot be calculated and thus cannot be predicted. On the other hand, it is very easy to calculate how often random events occur when considering larger numbers of cases: **probability** quantifies the likelihood/frequencies of such occurrences.

One example of the application of **probability theory** is modern climate research: without probability theory, there would be no reliable predictions in the fight against climate change and thus no solid basis for political decisions.

However, there were also scientists who did not want to calculate chance, but sought rather to eliminate it from their calculations if possible. For Einstein, for example, chance in quantum physics was a sign of its incompleteness. Yet he had to integrate it into his new photon theory.

It was not until the 1920s and 1930s that a coherent theory of probability was developed. The measure theory of the French mathematicians Émile Borel and Henri Lebesgue, developed from Cantor' set theory in the 1900s, served as a basis. The Soviet mathematician Andrei Kolmogorov, who like von Neumann in America made fundamental contributions to almost all subfields of mathematics and also directly influenced the most diverse fields of science, succeeded in 1933 in formulating the first contradiction-free theory. He showed that only three assumptions (axioms) are necessary for the description of a probability distribution: additivity, non-negativity, and normality.

Since, on closer examination, chance is involved in practically every event in the world, the way was now clear for a scientifically validated treatment of probability in the most diverse disciplines, from physics to economics and finance, from geology and weather research to sociology and historical research. Last but not least, the elevation of chance and probability to a scientific status made it possible for heredity to be deciphered in biology.

Measure theory, developed by Borel and Henri Lebesgue, is a branch of mathematics in which Leibniz' classical concept of the integral was further developed, and which today provides the foundation of probability theory.

As an example of the fact that chance can also lead to (almost) certain statements, Émile Borel devised the *Infinite Monkey Theorem*: If you let a monkey type long enough on a typewriter, it will eventually, with a probability of 1 ("0" means "impossible", "1" means: "will happen with a certainty of 100%"), type out all the books in the French national library. This thought experiment was "translated" into many languages. In England, for example, the monkey writes all of Shakespeare's sonnets.

The Birth of a New Science

By 1900, de Vries had rediscovered Mendel's work on heredity and linked it to the phenomena of variation and selection described by Darwin (see Chap. 4). In 1901 he published his mutation theory, which assumes, among other things, that there are smallest units of heredity that transmit the characteristics of a living being to its offspring according to Mendel's laws. This view was still controversial until the late 1920s. Among other things, many scientists assumed that a randomly occurring change in the genome would be distributed evenly among all offspring and thus attenuated beyond recognition. The fact that plant and animal breeders had completely different experiences was not taken into account in the academic world; there was hardly any exchange between scientists and non-scientists at the time.

Mutation: Spontaneous or deliberate change in the genetic make-up of an individual caused by external influences.

Biologists finally got to the bottom of heredity in a different way. The first question that arose was: Which cell structures are responsible for heredity? From the end of the nineteenth century, the resolving power of microscopes improved so much that details of plant and animal cells became more and more discernible. One of the many observations from this time were knobbly structures in the cell nucleus that became visible when the cell was preparing to divide (today we know that these structures are not visible in the light microscope before and after cell division because they then pass through the entire cell nucleus in an unwound form). Since these nodules could be easily stained, the German physiologist and anatomist Wilhelm Waldeyer-Hartz called them chromosomes in 1888 (*chrōma* is the Greek word for "colour", and *sōma* for "body"). Our knowledge of these strange structures in the cell nucleus soon increased:

- In all living beings, the number of chromosomes is constant, depending on the species. The pea has 14 chromosomes, the lamprey, a primitive fish species, 174, the mosquito 6, and humans 46.
- Chromosomes visible under the microscope look like squashed Xs at the beginning of cell division: two identical sister strands are held together at a constriction point.
- There is almost always a double set of chromosomes. In humans, there are 23 chromosomes of different sizes, each of which occurs in duplicate. Only the 23rd pair of chromosomes is different: one of the chromosomes is X-shaped, while in women the second one is also shaped like an X and in men it resembles a Y.

The process in which the female egg cell fuses with the male sperm cell could also be observed well with the microscope. While in all other cells the chromosomes are present in duplicate, in germ cells there is only a single set of chromosomes (in humans, 23). When egg and sperm cells fuse, the fertilised egg cell and each of the body cells of the offspring that later emerge from it (apart from the germ cells) have once again 46 chromosomes, 23 of them come from the father, 23 from the mother. This observation agreed well with Mendel's laws and was a strong indication that the chromosomes are indeed the structures responsible for heredity.

- In 1903, the American Walter Sutton and the German Theodor Boveri formulated the chromosome theory, according to which chromosomes are the carriers of hereditary traits.
- In 1906, William Bateson, one of Mendel's most prominent admirers, coined his own name for the science of heredity for the first time: genetics. This neologism was inspired by the ancient Greek: *geneá* means descent and *génesis* means origin. However, even Bateson could not believe in the role of the chromosome as the sole agent in heredity:

Much that is known of chromosomes seems inconsistent with the view that they are the sole effective instruments in heredity.[2]

● In 1909, the Dane Wilhelm Johannsen expanded the new vocabulary to include the name "gene" for the smallest unit for the transmission of a trait to the next generation. He called the full set of genes in an individual its genome.

> The reduction of the number of chromosomes in egg and sperm cells from 46 to 23 (in humans) takes place in **meiosis**. Put simply, in meiosis, a cell with 46 chromosomes divides into two cells, each with 23 chromosomes, without doubling the chromosomes. Which chromosomes in a germ cell originally come from the father and which from the mother is a matter of chance. There are 2^{23} (8.4 million) possibilities for the way the chromosome pairs are distributed between the two germ cells. If it is taken into account that both mother and father produce germ cells, there is a **probability of 1 in 35 trillion that** the germ cells of a parent pair will twice produce a fertilised egg cell with exactly the same chromosome distribution.

At that time, the chromosome theory was still hotly disputed. Many researchers considered it absurd that chemical compounds should be responsible for the characteristics of living beings. They thought that genes had to be abstract constructs. The American biologist Thomas Hunt Morgan belonged to this faction. He wanted to show experimentally that chromosomes were not gene carriers. But things turned out very differently than he had imagined.

Breakthrough in the Fly Lab

Because of the large number of characteristics of a living being, the equation could not possibly be "one chromosome—one characteristic". According to the chromosome theory, the chromosomes therefore had to be made up of smaller hereditary information units, presumably at fixed locations on the chromosomes. In heredity, therefore, certain traits would have to be passed on coupled together, and it was precisely this assumption that Morgan wanted to refute.

---

[2] W. Bateson, *Mendel's Principles of Heredity*, Cambridge University Press (1902), pp. 270–271.

In 1908, in search of a creature that is easy to keep, generates a large number of offspring in a short time, and passes on easily recognisable characteristics to the next generation, he came up with the idea of using the fruit fly *Drosophila melanogaster* as an experimental animal, thanks to a suggestion by his former doctoral student Nettie Stevens. This fly has only four pairs of chromosomes, produces around thirty generations a year and is easy to breed in empty milk bottles, using rotting bananas as food. Despite all these advantages, the work with *Drosophila* was anything but promising at first. Even after intensive irradiation and treatment with chemicals, Morgan could not detect any visible changes in the offspring of the experimental flies. He later categorised his experiments from this time into "foolish, downright foolish, and those that are worse".

It was not until 1910 that his wife discovered a specimen with white eyes instead of red. A mutation at last! Morgan had this white-eyed male produce offspring with red-eyed female flies. Of the 1237 offspring, all had red eyes. But in the following generation, out of 4252 flies, 798 specimens had white eyes!

- The result agreed with Mendel's observations in one important point: the white-eyed fly had passed on its specific trait to the *next but one* generation, according to Mendel's second rule.
- The ratio of (recessive) white-eyed to (dominant) red-eyed flies should have been 1:3 according to Mendel's rule. In this experiment, however, only about every fifth fly was white-eyed.
- Morgan noticed that all the white-eyed flies were male.

Nettie Stevens (1861–1912) was an outstanding scientist and an ace at the microscope. When she did her doctorate at Bryn Mawr College in Pennsylvania in 1903, she worked with *Drosophila* and mealworms. So Morgan Hunt, the head of the biology department and thus her supervisor, came up with the idea of also trying the fruit fly.

In 1905, in parallel with Edmund Wilson, Hunt's predecessor at Bryn Mawr College, Stevens discovered that the inheritance of sex is determined by the father's sperm cell: if it contains an X chromosome, the offspring becomes female; if it contains a Y chromosome, the offspring becomes male. (*Drosophila* and mealworms also have a larger X and a smaller Y chromosome).

The traits of an individual are thus not inherited independently, as Mendel had claimed, but in some cases coupled. Instead of an argument *against* the chromosome theory, Morgan had found one in its *favour*.

By the end of 1910, he had produced forty different, easily recognisable mutations: yellow wings, pink eyes, missing hairs, etc. The result of the first experiment with the white-eyed fly was confirmed: many mutations only occurred in certain combinations. For example, white eyes only occurred in combination with yellow wings, never with grey. Morgan concluded that these characteristics must be encoded on the same chromosome, so that both characteristics are inherited together.

But then one day he discovered a fly with white eyes and grey wings. The chromosome on which both characteristics are coded like pearls on a pearl necklace must have ruptured and the two fragments must have reassembled crosswise. It was an ingenious as well as correct conclusion: such a *crossing over* of chromosomes (subsequently referred to as *crossing over*) with subsequent exchange of small chromosome segments turned out to be a normal part of recombination. By recording the exchange values of various traits, Morgan and his colleagues were able to assign individual genes to the four chromosomes of Drosophila and produce ever more accurate gene maps.

Thus, it was finally discovered how changes in the genome of the offspring can occur repeatedly and purely by chance during sexual reproduction. Morgan published his findings in 1915 in his book *The Mechanism of Mendelian Heredity*. This and his later work *The Theory of the Gene* became early classics of genetic theory. In 1933, Morgan was awarded the Nobel Prize in Medicine and Physiology.

The ratio of 1:3 observed by Mendel in the transmission of traits to the next generation presupposes, among other things, that:
- the genes responsible for the observed traits are located on different chromosomes,
- *one* gene is responsible for *one* trait,
- there are no processes—such as crossing-over—that modify the genetic make-up of the offspring.

Mendel had randomly selected characteristics of the peas with which he came close to a 1:3 ratio. However, with a certain frequency, cross-overs must also have taken place. Therefore, it is assumed today that the report of his observations was somewhat "adapted" to the expected result.

The Unknown Compound in the Centrifuge

While physicists were busy with atomic energy, war technologies, the growing number of unknown elementary particles, and the still unsolved questions of interpretation of quantum mechanics, biology was heading towards a first high point in its history: the discovery of the mechanism of heredity. With the chromosomes, the scene of the action had been identified. What was still missing was the molecular composition of the genes. What did they consist of? Once again, it became apparent that an important missing piece of the puzzle had been lying around openly for decades; it was just that no one had been able to do anything with it until then.

In 1869, the Swiss researcher Friedrich Miescher studied the proteins contained in white blood cells. He used wound dressings from the local hospital as starting material. The gauze bandages soaked with pus were a rich source of this cell type. While trying to separate the proteins from each other, he noticed a previously unknown substance that had an astonishingly high phosphorus content and was definitely not a protein. Since it came from the cell nucleus, he called it nuclein. Miescher suspected that nuclein might have something to do with fertilisation and heredity, and he then examined the sperm of various vertebrates. However, he could not quite believe that *a single* chemical compound should control this crucial part of life. His colleagues took a similar view.

In the early years of the twentieth century, the Lithuanian–American biochemist Phoebus Levene began to study this substance, which had since been renamed nucleic acid, in more detail. Building on his experience and the results of other scientists, he was able to establish in 1929[3] that nucleic acid is a combination of three components:

- a sugar component that occurs as ribose or also as deoxyribose,
- phosphate,

---

[3] P. Levene, E. London: *The stucture of thymonucleic acid*. Journal of Biological Chemistry, 83, (1929). S. 793–802.

- organic nitrogen compounds in the form of the bases adenine, thymine, cytosine, and guanine.

Four of the five carbon atoms of the sugar moiety are joined together in a ring via an oxygen atom. This sugar portion in the nucleic acid can occur in two variations:
- In the first form, called ribose, the second carbon atom has a hydroxy group (OH group).
- In the second form, this OH group is missing, hence the name **deoxyribose**.

There are therefore two different types of nucleic acids: "ribonucleic acid" (RNA) and "deoxyribonucleic acid" (DNA). The missing OH group makes the nucleic acid strand of DNA significantly more stable than that of RNA.

Levene called the combination of a phosphate residue, a sugar, and an attached nitrogen base a nucleotide. The individual building blocks of nucleic acid were now known, but not their spatial structure. All that was known was that the individual nucleotides could form very long chains. Because the functioning of nucleic acid was also still mysterious, the view that it must be proteins that control heredity was not yet off the table.

The Molecule of Life

As in physics, Americans took the lead in biology worldwide from the 1940s onwards. In the early 1940s, for example, the American scientists George Beadle and Edward Tatum recognised how genes exert an influence on the characteristics of their carriers. They had induced mutations in moulds with X-rays and observed that in some cases they could no longer produce the enzymes necessary for growth. The two researchers concluded that the genes do not directly influence the characteristics, but that they produce and regulate the substances that are responsible for certain metabolic steps. The mutation of a gene can lead to the enzyme it regulates being defective or to its production ceasing altogether.

Beadle and Tatum formulated their insight that each gene controls the production of exactly one specific enzyme as the "one-gene-one-enzyme hypothesis". For this, they received the Nobel Prize in Medicine in 1958.

Today it is known that there are numerous exceptions to this rule. For example, the same chromosome section can cause the formation of different proteins.

It had been known, in fact since about 1930, that enzymes belonged to the class of substances known as proteins. These were therefore the molecules that carried out the formation of traits, but whether they were also the carriers of genetic information, as many researchers assumed, was questionable. At this point, nucleic acid moved back into scientific focus. The final proof that the seemingly unspectacular molecules of sugar, phosphates, and bases were actually the carriers of genetic information was achieved in 1944 by the Canadian Oswald Avery, together with the Canadian-born American Colin MacLeod and the American Maclyn McCarty. Their experiment was based on Griffith's experiments from 1928, in which he had shown that the genetic material of dead bacterial cells can be taken up by living bacteria and produce new characteristics there.

Hundreds of metabolic disorders are known in humans, most of which are very rare. Haemophilia, for example, is almost always caused by a genetic defect. Since a clotting factor is missing in the blood of those affected, wound closure is delayed.

In the meantime, it had been discovered that Griffith's experiment not only works with dead cells, but can also be carried out with cell extracts. But which substance in the cell extract transfers the genetic information? Avery and his colleagues treated cell extracts of the pathogenic S strain with various enzymes that specifically destroy carbohydrates, fats, proteins, and nucleic acids, and then injected them into the test animals. The result was clear: only the mice injected with the cell extract containing nucleic acid-destroying enzymes survived. In all other cell extracts, the DNA had remained intact and had been able to transfer the property of the S strain to the R strain. It was thus clear that the nucleic acid had to be the carrier of the genetic information.

Although Avery's experiments proved indisputably that nucleic acid must be the carrier of hereditary information, scientists doubted this result. The conviction that only the multiform proteins could be considered as carriers of the hereditary information was overpowering. Eight years later, in 1952, the American researcher Alfred Hershey provided further proof using radioactive markers. This time it was accepted by the scientific community that nucleic acid is indeed responsible for passing on genetic information to offspring. It is incomprehensible why Hershey received a Nobel Prize in 1969, while Avery, MacLeod, and McCarty went away empty-handed.

Griffith worked with two different pneumococcal strains in 1928. The individuals of the pathogenic **S strain** have a mucus membrane, so colonies of this strain look smooth and shiny—hence the name: s stands for smooth. The bacteria of the harmless **R mutant** lack the mucus capsule, and colonies of this type look rough in a petri dish—R stands for *rough*.

Griffith injected mice with R-type bacteria together with killed S-type bacteria. The mice died and live S-type bacteria were found in them. Griffith concluded:
- The dead S-bacteria must have contained genetic material in active form.
- This material was incorporated into the R cells so that the daughter generations also had living bacteria of the pathogenic type S.

The question of how nucleic acid manages to control countless characteristics and metabolic processes did not remain open for long. For it was precisely at this time that the new method of X-ray structural analysis made it possible to "see" the exact structure of the nucleic acid, and thus determine how it works.

The Discovery of the Double Helix

The Englishman James Watson finished school at 15, and by 19 he had degrees in philosophy and biology. This academic success in fast forward was accompanied by a huge ambition: he wanted to become famous. He started his scientific career in Copenhagen in the field of virus research. In 1951, at a conference in Naples, he met the New Zealander Maurice Wilkins, who told him about his attempts to crystallise DNA and examine its structure using X-rays. Watson decided to make a name for himself by decoding DNA, and in the same year, aged just 23, he moved to the Cavendish Laboratory in Cambridge. There he met the American Francis Crick, twelve years his senior, who as a physicist had been involved in the construction of sea mines during the Second World War and had switched to biology in 1947. When he met Crick, he had already been working on his doctoral topic for two years without much success: the X-ray structural analysis of haemoglobin. A deep friendship developed between Crick and the young Watson and they began to think about DNA—in their free time, because officially they both had other tasks. It was Maurice Wilkins who, as deputy head of the biophysics department at King's College in London, was working on decoding the DNA molecule.

In the early 1950s, not much was known about the DNA molecule:

- The individual building blocks of DNA each consist of a phosphate residue, a sugar, and a base.
- Sugar and phosphate molecules alternate as the "backbone" of the giant molecule, and the bases somehow attach to this chain.
- In 1952, Erwin Chargaff, who came from Austria and also emigrated to the USA in the 1930s, noticed that in the DNA of every living creature so far examined, the same amounts of the bases adenine and thymine are present, and likewise for cytosine and guanine (Chargaff rule).
- The American Linus Pauling, at that time the leading luminary in biochemistry, had developed a three-dimensional model for certain proteins in which their amino acids are arranged in a helix. He was considering something similar for the DNA molecule.

X-ray structure analysis: The substance to be examined, in crystalline form, is bombarded with X-rays. Because their wavelengths have the same order of magnitude as the atomic distances in the crystal, the crystal lattice generates diffraction patterns from which the structure of the substance can be read off. Today, computer programs help to interpret the resulting patterns.

X-ray images of crystallised DNA molecules by Rosalind Franklin, the specialist in X-ray structural analysis of crystallised macromolecules working at King's College in London, supported the assumption of a helix structure. But how many helices were there? One—as in the type of protein brought into play by Pauling? Or were there several interwoven strands? It was hard to imagine that such a multiple helix could be stable on the one hand and capable of twisting apart and copying itself in the process of cell division and reproduction on the other. And how would the nitrogen bases fit into the picture? Before their efforts were crowned with success, Watson and Crick had to overcome some setbacks:

- In 1952, Franklin presented her latest X-ray images of DNA in a lecture in London, and they showed clear evidence of a double-helix structure with inwardly oriented nitrogen bases. Watson was in the auditorium. Back in

Cambridge, he and Crick began to try out different arrangements of the various molecular groups of DNA with a construction kit system—small spheres represented the atoms (phosphorus, carbon, oxygen, hydrogen) and rods represented the chemical bonds between them. They quickly found a stable model and proudly presented it to Franklin and Wilkins. But Franklin immediately recognised the crucial error: the nitrogen bases in this model pointed outwards, not inwards as their pictures had shown. Watson had simply misremembered their lecture. They had to start all over again, with the nitrogen bases pointing *inwards* in the double helix structure.

- Crick and Watson would have liked to have a look at the latest results of Rosalind Franklin's X-ray structure analysis. But Franklin and Wilkins were not on good terms. Franklin had been hired by the head of the laboratory as an independent scientist with the task of investigating DNA using X-ray structure analysis. Wilkins, as deputy head, thought Franklin was his assistant. The organisational confusion led to Franklin guarding her data sets like a treasure and not wanting to share them with anyone—certainly not with Watson and Crick, who worked at a completely different institute.

- Watson and Crick had overlooked or failed to interpret Chargaff's publication, which showed that the four bases always come together in the same pairs. They came to this realisation by a different route. Crick asked a mathematician friend, John Griffith, if he could use mathematics to determine what possibilities there were for juxtaposing the different bases. Griffith found out that there were only two possibilities: the pairing of adenine with thymine and of cytosine with guanine, exactly what Chargaff had long since found out. Valuable time had been lost! Chargaff confirmed this finding in a conversation with Watson and Crick in 1952. However, the three did not get on well together. Chargaff saw in Watson and Crick only ignorant beginners and "scientific clowns".

- On top of all this, the two researchers received deeply sobering news from America. They knew that Pauling had also been working on decoding the DNA structure in America. Now they heard that he had completed a model of the molecule. The race seemed lost.

Proteins are strings of amino acids. Joined by hydrogen bridges, the amino acid chains take on a three-dimensional, folded structure.

In this seemingly hopeless situation, the tide suddenly turned. Watson and Crick had befriended Pauling's son Peter, who happened to be in Cambridge as a young researcher. From him they received a preprint of Pauling's work. They could hardly believe it: he had made the same mistake as they had in their first model! He too had placed the nitrogen bases on the outside. So the race was back on again. Watson and Crick decided to appeal to Wilkins once more and get him to cooperate. The latter told Watson and Crick that Franklin had seen even clearer structures of the double helix structure in more recent pictures. Again, she had kept the pictures to herself. Wilkins had no choice but to secretly make prints of Franklin's results (in other words, steal them) and share them with Watson and Crick. With this crucial clue, the two returned to their DNA kit. After five weeks of tinkering, their model was finally coherent. It was wonderfully simple and at the same time intuitively plausible:

- DNA consists of two helical strands of nucleotide chains wound together like a spiral staircase. The outside boards are the two sugar-phosphate chains, and the steps are the nitrogen base pairs connected by relatively weak hydrogen bonds.
- During cell division, these hydrogen bonds break and the double helix opens like a zip. One helix at a time is supplemented by cell contents, so that at the end of the reproduction process there are two identical double helices.

The publication of Watson and Crick's 1953 paper in which they presented the universal structure of DNA is a high point in the history of biology and the sciences as a whole. This structure also provided the first clues as to how DNA works. Watson and Crick wrote in their epoch-making paper:

It has not escaped our notice that the specific pairing we have postulated immediately suggests a possible copying mechanism for the genetic material.[4]

In this way, they anticipated the finding that the arrangement of nitrogen bases encodes the production of a wide variety of proteins.

- In 1962, the two were awarded the Nobel Prize for Medicine together with Wilkins.

---

[4] J. Watson and F. Crick: Molecular *Structure of Nucleic Acids: A Structure for Deoxyribose Nucleic Acid*, Nature, vol. 171, pp. 737–738 (25 April 1953).

- Erwin Chargaff was not considered for the Nobel Prize in 1962 for the discovery of DNA structure. He bitterly withdrew from active scientific life.
- Linus Pauling was narrowly defeated in the race to decode the atomic structure of life. Nevertheless, he did not go away empty-handed that year: after receiving his first Nobel Prize in 1954 for his theory of chemical bonding, he was honoured with the Nobel Peace Prize in 1962 for his commitment against nuclear testing.
- Rosalind Franklin, who had played a decisive role in the development of Watson and Crick's model with her X-ray structural analyses, had died of cancer in 1958 at the age of 37. For a long time, Watson and Crick denied that they had seen the DNA images stolen by Wilkins. Only in his book "The Double Helix" from 1968[5] did Watson admit to having had access to the data.

From DNA to Proteins

Once the DNA structure was known, the next question was: How does this giant molecule, which is far too large to pass through the wall of the cell nucleus, encode the formation of proteins that takes place outside the cell nucleus? In painstaking detail, the researchers were able to show that the other type of nucleic acid, ribonucleic acid (RNA), plays a decisive role here.

- In the cell nucleus, individual sections of the DNA are made accessible so that the bases are exposed and an RNA with complementary bases can form along them—if there is a guanine on the DNA, a cytosine is incorporated into the RNA, and so on. (There is only one difference: instead of thymine, RNA works with a base called uracil, which has very similar properties.) The RNA is now a mirror image of the DNA segment. Since it is single-stranded and also only forms shorter chains, it can make its way through the wall of the cell nucleus and move around in the cell to the places where protein synthesis takes place.
- Here, it serves as a blueprint for the structure of the protein that the DNA section codes for. With the help of another type of RNA, amino acids are strung together in a chain according to the base code, which folds at the specified points and forms a three-dimensional structure that gives the resulting protein its unique nature and function.

---

[5] James Watson, *The Double Helix – A Personal Account of the Discovery of the DNA Structure*, Atheneum (1980), (Original: 1968).

- In each case, a specific triple combination of RNA bases is associated with exactly one amino acid. This code applies to all living organisms with a few exceptions.

In 1961, Crick proved that three consecutive bases each encode exactly one amino acid. The American biologist Marshall Warren Nirenberg gave the first answer in 1961. He proved that the sequence uracil—uracil—uracil encodes the amino acid phenylalanine. He also succeeded in deciphering the code for the amino acids serine and leucine. The assignment of the other amino acids to their triplets was a demanding piece of work, much of which was done by the Indian chemist Har Gobind Khorana. Nirenberg and Khorana shared the Nobel Prize for Medicine in 1968.

> The 21 amino acids cannot be uniquely coded with base pairs, because only two of the four available bases would result in only 16 possible combinations. Triplets with four possible building blocks each result in 64 possible triple combinations. Thanks to this redundancy, certain amino acids are represented by several different triplets.

It is a miracle of nature: the structure of DNA and RNA is simple and requires very few building blocks, but the biochemical processes of heredity and protein synthesis associated with them are extremely complex. Many details were only understood decades after the discovery of DNA and RNA, and others are still the subject of research.

The knowledge of biochemical control processes has led to the emergence of the new science of genetic engineering, and triggered a wave of new technical possibilities. Modifications of genes open up many possibilities for genetic design.

- **"Red" genetic engineering**: With the help of genetic engineering processes, active substances can be produced for drugs and other substances needed in medicine. In 1982, the first genetically engineered drug, human insulin, came onto the German market. The sought-after substance no longer had to be extracted from millions of pig pancreases, but was now produced by bacteria. Today, every second new drug in this country is now produced by genetic engineering. Last but not least, thanks to genetic engineering, it was possible in the coronavirus crisis of 2020/21 to develop a very effective vaccine against the virus within just a few weeks.
- **"White" genetic engineering**: In many industrial sectors, conventional chemical processes are increasingly being replaced by genetic engineering

methods, because biological processes consume less raw materials, energy, and water and generate fewer waste products than chemical processes. The production of new groups of substances is also possible through genetic engineering; for example, biopolymers similar to spider silk are used in medicine and in the cosmetics and textile industries.

However, the targeted genetic manipulation of living beings (and viruses) is also associated with very real risks, not least because the genetic material of humans can also be changed. The discussion about whether and how genetically modified organisms can be prevented from getting out of control, and how human dignity can be protected (think, for example, of those who might want babies with desired characteristics) will accompany us for many years to come.

11

The Pyrrhic Victory of Big Science—How Science Was Domesticated by the Military and Industry

By the middle of the twentieth century, the sciences had survived their fundamental crises and were on their way to reshaping the world as never before.

- **Quantum physics** had replaced Newtonian classical physics. Although its basis, individual quanta, defied calculation, it achieved an explanatory power and accuracy in predicting events that was unparalleled in the old physics. The leap from academic knowledge to technology generated a multitude of applications that would have been unimaginable to earlier generations.
- A similarly paradoxical development had taken place in the field of **mathematics**. It had detached itself from anything previously imaginable, and it was precisely through this tremendous degree of abstraction that it had opened doors to new knowledge and their associated technologies.
- The science of **biology** had suddenly outgrown its main activity of collecting and cataloguing and had turned to the mysteries of metabolism (physiology) and heredity (genetics). The bookkeepers of life had become actors. The final breakthrough was to decode the composition and function of DNA.

Fundamental research rushed from success to success, and our knowledge increased in leaps and bounds. However, the transformation of this knowledge into technologies would now take place under completely different

L. Jaeger, *The Stumbling Progress of 20th Century Science*, https://doi.org/10.1007/978-3-031-09618-1_11

circumstances. In the nineteenth century, the main drivers of technological development were trade, transport, and human health and comfort (although these three factors are closely linked). But with the rearmament that preceded the First World War, there came a turning point: the creative power of scientists and the direction of technological progress were appropriated by the military—the development of poison gas by Fritz Haber is an early example. From the 1930s onwards, the transformation was almost complete: technological progress now paid off almost without exception in military applications. The four most important technologies of the day were relentlessly pushed forward before and during the Second World War. It was only much later, as a kind of afterthought, that they found civilian applications and began to play a role in our everyday lives.

Four **technologies** that were developed in the course of the Second World War:

1. Nuclear fission
2. Radar
3. Rocket propulsion
4. Production of penicillin.

1. **Nuclear fission**: The development of the atomic bomb was only possible because scientists and technicians from all over the world worked together on a project in an unprecedented way. As far as circumstances allowed, every restriction, especially in terms of financial resources and technical equipment, was removed by the military.

   The atomic bomb and the hydrogen bomb, built only a few years later, continued to be the main drivers of nuclear technology after 1945. Soon the Soviet Union, England, France, and from 1964 China were also official nuclear powers. The emerging balance of terror devoured hundreds of billions of dollars every year.

   - In 1950, the USA possessed 379 nuclear weapons, the USSR 5.
   - In 1986, at the beginning of nuclear disarmament, the USA had 23,250 nuclear weapons, the USSR 40,700.

It took six and a half years from the first nuclear fission in December 1938 by Lise Meitner and Otto Hahn in Berlin to the explosion of the first atomic bomb in the Arizona desert in July 1945. Another six and a half long years passed before the peaceful use of nuclear energy began.

- In December 1951, an experimental reactor in the USA began to generate electricity for the first time—enough to light four light bulbs.
- In 1953, US President Eisenhower, the former commander-in-chief of the Allied forces in Europe, delivered his "Atoms for Peace" speech to the UN, in which he held out the prospect of disseminating the knowledge accumulated in the USA about nuclear power:

So my country's purpose is to help us to move out of the dark chamber of horrors into the light, to find a way by which the minds of men, the hopes of men, the souls of men everywhere, can move forward towards peace and happiness and well-being.[1]

- In 1954, the world's first nuclear power plant went into operation in the Soviet Union. By 1977, a total of 200 nuclear power plants were delivering electricity; from 1989 onwards (420 nuclear power plants), the trend slowed down considerably; in 2020, about ten percent of the world's electricity production came from 442 nuclear power plants.
- In 1955, the first atomic energy conference took place in Geneva. Delegates held intense discussions about the possibilities, but also the dangers, of nuclear fission.

Hydrogen bomb (H-bomb): While an atomic bomb draws its energy from nuclear fission, nuclear fusion takes place in H-bombs. The enormously high pressure that causes atomic nuclei to fuse is generated in the H-bomb by an atomic bomb. They are therefore two-stage, the atomic explosive is the detonator for the H-bomb. Their destructive power is greater than that of the atomic bomb alone by a factor of 100–1000.

The debate about the advantages and disadvantages of nuclear energy has never reached a conclusion. It has been particularly intense again recently because of the need to supply energy to the world without increasing CO_2 emissions.

However, atomic fission was only one of the many technical possibilities that arose from quantum physics. The military's pursuit of the construction of atomic and hydrogen bombs fuelled knowledge about the quantum world. As

---

[1] From Dwight D. Eisenhower's speech to the UN General Assembly in New York on 8 December 1953, https://www.iaea.org/about/history/atoms-for-peace-speech.

a result, research into other technologies was also intensified after the Second World War.

2. **Radar**: At the beginning of the Second World War, the development of radar was more advanced in Germany than in any other country. On the ground and in the air, this new technology was used to defend against Allied bombers. But since the German leadership believed the war was as good as won, research in this field was largely halted from the end of 1940. Great Britain and the United States, on the other hand, intensified their radar research. As early as the second half of 1940, the British were able to locate German aircraft in time to prevent the air supremacy sought by Germany and thus also the invasion of Great Britain. By the end of January 1943, the technology had advanced so far that British planes were able to make out the contours of the landscape by radar for the first time during a bombing raid on Hamburg. The devastating bombing raids of the following period led to the realisation in Germany that radar research had been stopped too early. But it was too late to catch up.

Radar stands for *radio detection and ranging,* i.e., radiation-based tracking of the desired items and perform distance measurement.

This is how a radar works. Electromagnetic waves are emitted from a transmitter. When the signals hit an object, they are reflected and can be detected by a receiver. From the properties of the reflected waves, information can be obtained about the object that reflected them, including its distance and speed. The transmitter developed by the British was particularly powerful: thanks to the very short wavelength of the signal, the resolution was so great that it was even possible to recognise the contours of an aircraft and thus its type. The technical name of the small, palm-sized transmitter—the magnetron—is almost unknown outside the technical world today, but what it generates is firmly anchored in our everyday lives: microwaves.

- Microwaves have been used in the kitchen since the 1960s.
- In organic chemistry, microwaves with pressure and temperature control are used to synthesise certain products, including drugs such as acetylsalicylic acid (aspirin) and flavourings such as vanillin.
- Among the many other technical applications, a good example is the decontamination of contaminated soil. The soil is heated with microwaves

to the boiling point of the harmful organic compounds and the escaping gases are collected and disposed of.

- **Rocket technology**: Germany remained the leader in this field until the end of the Second World War. The Peenemünde testing facility was the largest military research centre in Europe from 1936 to 1945. The V2 rocket (V stands for Vergeltungswaffe, Weapon od revanche) built here was the world's first guided long-range missile. The tip of the 14-m-long V2 had room for a tonne of explosives. Powered by a mixture of oxygen and alcohol, it reached an altitude of over 84 kms, making it the first man-made object to reach the edge of space. By the end of March 1945, over three thousand rockets had been fired at targets in England, Belgium and France, with over a thousand hitting London. The wonder weapon was not very accurate, however, and its effect was more psychological than anything else: since it flew at almost five times the speed of sound, there was no possibility of defending against it. Many contemporary witnesses reported the disturbing fact that such a missile could strike anywhere, at any time, and without any warning.

After 1945, the scientists and engineers involved in the design and production of missiles left their home countries, sometimes voluntarily, sometimes not, for the USA or the Soviet Union. During the Cold War, the technology of intercontinental weapon carriers became a fiercely contested discipline between the Eastern and Western blocs.

There are different ways to define the limit between the atmosphere and space.
- The US Air Force and NASA assume that the limit is around 80 kms, because above this altitude the air pressure becomes so low that an aircraft can no longer be *controlled aerodynamically.*
- In the 1950s, the Hungarian–American physicist Kármán defined the transition to space differently: above a certain altitude, an aeroplane would have to fly so fast in order to still have sufficient *aerodynamic lift* that centrifugal forces would propel it into space. The height of this Kármán line is 100 kms.

The technical director of the Peenemünde Army Experimental Station and self-confessed Nazi Wernher von Braun made a career in the USA. He was the indispensable architect for the intercontinental missiles needed in the Cold War. But the development of the Saturn rockets for civilian use, an important part of the Apollo programme with the goal of landing on the Moon, also took place under his leadership.

The only way to get into space is with rocket propulsion; every satellite orbiting the Earth has at some point been launched into orbit by a rocket. In 2021, there were about 4500 satellites in space, and the ESA estimates that this number will increase to about 50,000 by 2030. How many of today's satellites serve partly or exclusively for military purposes is, of course, unknown; estimates put the number at three to four hundred. But one thing is certain: without (civilian) satellites, global communication, GPS navigation, weather observation, and much more would collapse.

4. **Production of penicillin**: In addition to physics, medicine and biology also contributed to the development of technologies that initially determined the outcome of the Second World War and were only later used for civilian purposes. In the First World War, more soldiers had died from infections than from direct weapon impact, so the US military supported the mass production of penicillin, which had been rediscovered in the early 1940s, with all the means at its disposal. In time for D-Day, the planned invasion of Normandy on 6 June 1944, 2.3 million doses of the antibiotic were available to the Allied troops. Now it was possible for the Allied invasion forces to mitigate infections in soldiers and avoid major casualties by administering penicillin. This had an impact on the outcome of the Second World War similar to that of the atomic bomb, radar, and rockets.

In the first few years, the distribution of this new miracle drug was limited to soldiers; for civilians it was almost impossible to get penicillin. In Germany, this shortage lasted until 1950; only a few people could afford the black market prices. Given the urgency of the situation, German researchers came up with the idea of collecting the urine of soldiers treated with penicillin in Allied hospitals. This is because a large part of the administered antibiotic is excreted by the body again within a few hours—this is one of the reasons why the treatment has to be repeated regularly for ten days or longer. Penicillin extracted from urine saved the lives of many Germans. From 1950 onwards, the original production method using moulds gained momentum in Germany.

The industrial production of penicillin was one of the most important advances in medicine of the twentieth century; it freed people from the grip of numerous often fatal diseases. The history of its discovery and development as a medicine will therefore be described in more detail in the following pages.

Penicillin—The Miracle Drug of Modern Medicine

Until well into the middle of the twentieth century, millions of people died each year from bacterial infections such as pneumonia, meningitis, and wound sepsis. Some diseases could be held at bay with vaccinations. For example, prophylaxis had been possible against diphtheria since 1923. This is a bacterial infection of the throat and pharynx and was the most frequent cause of death among small children in the nineteenth century. But when a mass reproduction of bacteria had already begun, there was no effective medicine. The search for a substance that could treat infections was correspondingly intensive. In the 1920s and 1930s, many companies were founded in Europe and America that had recognised the tremendous business opportunities of industrialising chemistry. The most prominent of these were the Merck Company in America and the German I.G. Farben in Europe (which became Bayer, BASF, and Hoechst after the Second World War). The laboratories and workshops of these companies produced.

- medication,
- new materials such as Bakelite and rubber,
- mass consumer goods like nylon stockings.

Despite the systematic research carried out, the mass production of the world's first antibiotic that worked reliably and had few undesirable side-effects can be traced back to a coincidence. When the Scottish biologist Alexander Fleming returned to the laboratory from his summer holiday in 1928, he discovered some forgotten Petri dishes on the table, containing dangerous staphylococci whose colonies had in the meantime overgrown the entire culture medium. When he went to clean the glass dishes, he noticed that mould had grown in some of them. Such contamination was not uncommon, but there was a clearly visible zone around the mould colonies that was free of bacteria. Some substance released by the mould into the nutrient substrate must have prevented the bacteria from colonising these areas as well.

The fact that moulds can be antagonists of bacteria had long been known:

- As early as the Middle Ages, doctors used moulds specifically to treat wound infections, without knowing why this therapy sometimes worked. It was a gamble for their patients, because not all mould species produce penicillin.

- From the second half of the nineteenth century, several researchers described the so-called antibiosis of mould and bacteria. But their experiments showed that what kills the bacteria also severely damages the infested organism.

Fleming identified the mould in his Petri dishes as a species of the genus *Penicillium*, which was rather rare at the time. He initially had difficulty replicating his observation because he made the mistake of adding the fungus to culture media already overgrown with staphylococci, expecting the growing mould to eat holes in the bacterial lawns. Penicillium, however, only *inhibits* the growth of bacteria; it does not affect already existing bacterial colonies. Only when Fleming cultivated fungus and bacteria in parallel was he able to reproduce the observation that the bacteria did not grow into the zones around the fungal colonies.

The biggest problem proved to be the task of isolating the active ingredient penicillin. After many failures, Fleming began to wonder whether a remedy against bacterial infections could ever be obtained from the mould. He published the results of his experiments in 1929, pointing to the possible therapeutic application.

Despite his doubts, Fleming never gave up his work with penicillium. Another setback was that a completely different medicine against infections was being developed at the same time with great success. In Germany, the physician Gerhard Domagk had been researching bacteria-inhibiting substances on behalf of IG Farben from 1929. He recognised the potential of certain sulfonamide compounds, which were actually produced in-house as a basis for dyes. In 1935, the sulfonamide Prontosil, produced by Bayer-Werke, came on the market and was used against inflammations, blood poisoning, childbed fever, and many other diseases. Despite this significant breakthrough in the fight against infectious diseases, Fleming was convinced that his penicillin was even better. After a lecture given by Domagk to the Royal Medical Society in late 1935, Fleming approached him and said:

I've got something much better than Prontosil, but no one'll listen to me.[2]

---

[2] https://www.forschung-leben.ch/publikationen/biofokus/penicillin-und-sulfonamide-im-kampf-gegen-infektionen-zwischen-begeisterung-und-skepsis/ (Original in German).

The fact that **Alexander Fleming** found a mould culture that affected certain bacteria was due to chance and, moreover, was not a new discovery—it had long been known that some moulds have an inhibitory effect on bacteria. His real feat was knowing that the substance secreted by this particular Penicillium species harms bacteria but is not toxic to humans or animals.

Once again, the Germans made the wrong decision. Believing that the discovery of synthetically produced sulphonamides would do what was needed, they left the field of penicillins from natural sources to England and America. By the end of the 1930s, the effectiveness of sulphonamides had already begun to wane as resistant strains of bacteria began to appear. At that time, England and America had long since achieved their first successes on the way to the large-scale production of penicillin.

Ten years after Fleming's discovery, in 1938, the Australian Howard Florey and Ernst Chain, who had emigrated from Germany to England, came across Fleming's work in their search for bacteria-inhibiting substances. In painstaking laboratory work, they succeeded in producing pure penicillin and also proved that this substance could be used to combat, not only staphylococci, but also other aggressive strains of bacteria. They managed to produce so little penicillin that they could treat just one person, a 43-year-old policeman.

The British pharmaceutical companies asked by Chain and Florey for cooperation saw no possibility of bringing penicillin to the market on an industrial scale, because the production of penicillin was extremely costly. However, in America, a much larger country, penicillin finally began its triumphal march, not least because of media interest in the story of another person cured by penicillin. The American Anne Miller had suffered blood poisoning and had been close to death, but was saved by penicillin. The military importance of penicillin was obvious, so in 1943 the American War Production Board (WPB) took control of the mass production of this new antibiotic. Out of over 175 companies, it selected 21 to work on the penicillin programme—including Merck and Pfizer. Even though the WPB ensured the supply of all kinds of resources in the middle of the war, the production of penicillin remained an extreme challenge. Pfizer's vice-president, John L. Smith, is said to have exclaimed:

The mould is as temperamental as an opera singer, the yields are low, the isolation is difficult, the extraction is murder, the purification invites disaster and the assay is unsatisfactory.

The London policeman Albert Alexander was the first person to be treated with **penicillin**. In 1941, he had scratched the corner of his mouth in an everyday accident; the tiny wound became infected and he developed blood poisoning. Following the treatment with penicillin, he seemed to make a full recovery. But the story does not have a happy ending. It was not yet known that treatment with penicillin had to be given for at least ten days, even if the disease seemed to have subsided. Nor was there that much material available at the time. The symptoms of blood poisoning struck again and Alexander died a few weeks later.

But the Americans succeeded. In 1943, the first field trials with penicillin took place in the fiercely contested region of North Africa, and its effect exceeded all expectations. The antibiotic became the most important weapon against bacterial diseases and inflammations of all kinds, and shifted the balance of power between the warring parties in favour of the Allies. For the discovery (and rediscovery) of penicillin, Fleming, Chain, and Florey were jointly awarded the Nobel Prize for Medicine in 1945.

The triumph of penicillin also revolutionised the history of pharmacy in another way. For a long time, it had been considered that only synthetically produced agents such as sulphonamides could be successful against bacteria; the extraction of natural substances from living organisms seemed technically too unsafe for scientists. But now researchers set out to find other natural sources of antibiotics. In 1946, Selman Waksman, who came from Kiev and emigrated to the USA, and his doctoral student Albert Schatz at Rutgers University discovered a remedy against tuberculosis bacteria, one of the most dangerous and deadly types of bacteria at the time: this was streptomycin, which originated from the bacterium *Streptomyces*. For this, Waksman (but not Schatz, who was instrumental in the discovery) received the Nobel Prize for Medicine in 1952.

The American company Merck played a major role in the discovery and development of streptomycin.

- It supported Waksman's research both financially and with staff in its own laboratories.
- It built a factory in Virginia to produce streptomycin.
- Its lawyers managed to secure patents on the new antibiotic—it had been disputed whether the drug was a non-patentable natural substance or a synthetic product.

> Selman Waksman coined the name given to the metabolic products formed in fungi or bacteria that either kill or inhibit pathogens: **antibiotics**. Today, synthetic drugs that inhibit bacterial infestation are also classified in this group.

When it became clear that Merck alone could not satisfy the great demand for streptomycin, it transferred the rights to Rutgers University at Waksman's insistence. Now other companies could also acquire licences and produce the new cure. Noble gestures like this are also possible in modern times: when the British–Swedish company Astra-Zeneca brought a coronavirus vaccine to the market, it made it public that it wanted to market the vaccine at cost price. In the first quarter of 2021 alone, global sales of the vaccine amounted to $275 million, with profits of $0 as promised.

Quantum Effects in Everyday Life

Just like radar, rocket technology, and antibiotic development, the technological implementation of quantum theory was originally used for military purposes in the Second World War—the result was the atomic bombs that abruptly ended the war with Japan. Only later did applications of quantum physics find their way into civilian life.

When physicists developed the abstract mathematics of the wave function and quantum statistics in the 1930s and 1940s, they realised that some quantum effects can also show up directly in the macroscopic, directly observable world. It is precisely these phenomena that lend themselves to human use. They include:

1. Emission spectra of the elements
2. Superconductivity
3. Superfluidity
4. Lasers.

1. The **emission spectra of the elements** had already been known since 1802 and became an important tool in the analysis of chemical compounds. The fact that it is ultimately a quantum phenomenon only became clear with the establishment of quantum theories.
2. **Superconductivity**. In some materials, their electrical resistance disappears as soon as they are cooled down to a certain temperature that depends on their composition. In this state, they can conduct an electric

current without losses; even with very high currents, the material doesn't heat up at all.

- In the case of aluminium, the temperature at which superconductivity occurs is −272 °C, i.e., about 1 °C above absolute zero.
- For mercury, the value is −264 °C.
- In 1986, the Swiss physicist Alexander Müller and the German mineralogist Georg Bednorz discovered a ceramic material that already becomes superconducting at −238 °C. Less than a year after publishing their discovery, they were honoured with the Nobel Prize.
- In 2020, the temperature record for a superconducting material was +15 °C The structure of the material is no longer "pacified" here by low temperatures, but by holding it at extremely high pressure. In this case, 267 gigapascals were necessary, where 1 gigapascal corresponds to 10,000 times the pressure of the Earth's atmosphere. For comparison, at about 6 gigapascals, graphite is transformed into diamond.

According to the laws of classical physics, the free electrons of an electric circuit constantly interact with the atoms of the conducting material as they travel through it. With each contact, a certain amount of energy is radiated, for example in the form of heat. The Dutch physicist Heike Kamerlingh Onnes was the first to describe this phenomenon. For a long time, he and his colleagues were unable to explain the fact that there is current transport without resistance losses. It was only quantum theory that allowed physicists to understand why there is no loss of energy of the electrons and thus no electrical resistance in superconducting materials:

- The atomic nuclei of a conductive material form a lattice through which free electrons move when a voltage is applied. At room temperature, the atomic nuclei in the lattice vibrate so strongly that they impede the flow of electrons.
- At very low temperatures (or if previously confined at very high pressures), the oscillations of the positively charged atomic nuclei in the lattice are much reduced. Now certain quantum processes cause the positive charges of the atomic nuclei to slide. As a result, the electrons experience a stronger attraction from the atomic nuclei than the repulsion between their own negative charges. This causes them to join together to form so-called Cooper pairs.
- Free electrons each have a spin of $+1/2$ or $−1/2$ and therefore obey Pauli's exclusion principle as fermions. A Cooper pair, on the other hand, has a

total spin of zero and behaves like a boson, for which Pauli's principle does not apply.

- All bosonic electron pairs of the electric circuit are connected in a single wave function and are in a common, macroscopic quantum state; when voltage is applied, they move through the lattice like a *single* particle.
- If a single Cooper pair were attracted to a positively charged atomic nucleus of the conducting material and released energy, the same would have to happen to all the other Cooper pairs at the same time. The local electrical forces of the lattice of atomic nuclei fall far short of being able to do this. The system is therefore *forced* to suppress any energy transfer from individual Cooper pairs to the lattice. Thus, the current flows without resistance.

This quantum mechanical process was deciphered in 1957 by the US physicists John Bardeen, Leon Cooper, and John Schrieffer, who were awarded the Nobel Prize in Physics for their BCS theory in 1972.

Superconductors have long since found their way into everyday civilian life. Above all, their ability to generate very strong, stable, and large-scale magnetic fields is used:

- in magnetic resonance imaging (MRI), also known as nuclear magnetic resonance (NMR) imaging, which has become an indispensable imaging technique in medicine,
- in mass spectrometers and particle accelerators,
- in test facilities for nuclear fusion reactors, which could revolutionise energy supply in the future,
- in power supply networks.

Superconductivity in electricity supply: Since 2014, an approximately one-kilometre-long cable cooled with liquid nitrogen has been running under Essen's inner city. The superconducting 10-kilovolt cable replaces a conventional 110-kilovolt line and reliably supplies customers such as department stores and banks with electricity. Based on the success of this trial operation, the Munich city works are planning to build a 12-km superconducting power cable.

Since superconductors will float in a magnetic field, it is also conceivable to build a magnetic levitation train, which would enable almost resistance-free and cost-effective transport.

3. **Superfluidity:** The macroscopically observable quantum state of liquid helium, whose temperature falls below a very low critical value, was discovered in 1938 by the Soviet physicist Pyotr Leonidovich Kapitsa and independently by the Canadians John F. Allen and Don Misener. The phenomenon contradicts our everyday intuition:

- Since fluids in the state of superfluidity have no internal friction, a vortex generated in them will persist almost indefinitely.
- They flow without resistance through the narrowest capillaries and even creep up the vessel walls when a level needs to be equalised.
- They have a high thermal conductivity. Superfluid helium conducts heat 30 times better than copper.

This macroscopic quantum effect also occurs because a large number of bosons occupy a common quantum state. So far, superfluidity has only been observed in helium and a lithium isotope. Superfluid helium is used in spectroscopy, including NMR spectroscopy.

4. **Lasers:** The laser is the only phenomenon presented here that was not first observed in an experiment and later explained theoretically. Here, the discovery went exactly the other way round. The fact that electrons can jump from one energy level to another during their movements around the atomic nucleus had been recognised by Einstein in 1905 and Bohr in 1913. There are two possibilities for these proverbial quantum leaps:

- **Absorption of a photon:** By absorbing the energy of an incoming photon, an electron jumps from a lower energy level to a higher one. The photon "disappears".
- **Emission of a photon:** An electron falls spontaneously from a higher energy level to a lower one; the energy released in the process is transferred to the electromagnetic field and produces a photon, which is emitted.

But how can these processes be put into mathematical form? A quantum theory for light did not yet exist at that time. From 1917 onwards, Einstein devoted himself once again to the phenomenon of light absorption and emission. He was a thinker, not so much an experimenter, and through theoretical considerations alone, he succeeded in mathematically describing the fate of photons spontaneously absorbed or emitted in the atom. In the process, he tracked down another possibility for photon emission:

- **Induced emission:** An incident photon causes the electron to leave its energy level. This time, however, it falls to an even lower level and this quantum leap creates a new photon. Together with the photon that triggered the effect and remains unchanged, *two* photons now leave the atom. While spontaneous photon absorption and emission are processes that can take place spontaneously at any time and cannot be individually predicted, induced emission is a causal and thus controllable process.

In an environment where many atoms are in an excited state and correspondingly many electrons are already at a higher energy level, the induced emission can set off a chain reaction of electron jumps to lower levels and thus an avalanche effect of emerging photons. What is special about this is that, just like the electron pairs in superconductivity and the helium atoms in superfluidity, all the emitted photons have exactly the same properties. They oscillate in the same phase, propagate in the same direction, and have the same frequency and polarisation—physicists speak of a coherent light wave.

A few photons that set a chain reaction in motion produce an intense visible, light whose photons have identical properties. This effect is called *light amplification by stimulated emission of radiation,* or **laser** for short. Laser technology was a key quantum technology of the twentieth century and is expected to be superseded in importance by the quantum computer.

When experimental proof of laser radiation was achieved in the 1950s and 1960s, it paved the way for technological implementation. Today, the laser is the best-known and most widespread macroscopic quantum effect and plays an important role in many areas of life:

- In medicine, it has become indispensable for cancer diagnostics, the treatment of kidney stones, and the correction of eye lens curvature, among other things.
- Science uses laser-based devices for spectroscopy, pollutant measurement, speed determination, and earthquake prediction, for example.
- In communications, lasers are used for fibre optic communication, underwater communication networks, and space communication.
- In industry and technology, laser beams are used to cut glass and quartz, burn ultra-small integrated circuits, read barcodes, drill aerosol nozzles and control elements with high precision, retrieve stored information from compact discs (CDs), store large amounts of data on CD-ROMs, print paper, and much more. Should energy be obtained from nuclear fusion in the future, the laser will be indispensable there as well.

- In the military, the laser is used, among other things, in the form of laser weapons, for range finding, and for reconnaissance at night.

A Strong Team: Quantum Theory, Logic, and Computer Science

The computer is another example of how abstract and non-descript quantum theory can nevertheless produce very concrete technological applications.

A major problem of early computer design was the size and fragility of the glass vacuum tubes (electron tubes) that were used to amplify and modulate electrical signals until the 1950s. In the search for a more stable alternative, engineers came across the class of materials known as semiconductors. Their conductivity lies between that of good electrical conductors (such as most metals) and non-conductors (such as plastic, glass, or porcelain).

On 23 December 1947, the Americans John Bardeen, William Shockley, and Walter Brattain succeeded for the first time in controlling current flows in a system of semiconductors. The building blocks made of semiconductors, known as transistors, proved to be far superior to vacuum tubes.

- They can be very finely controlled by external parameters such as voltage, temperature, or the introduction of foreign atoms into their crystal structure (doping).
- The first transistors were only 1/50 the size of vacuum tubes. And so began the miniaturisation of computers.

The discovery of the transistor effect paved the way for realising von Neumann's and Turing's dream of handy and super-fast calculating machines. The transistors became smaller and smaller, and from the 1970s they were packed together on microprocessors and memory chips to form integrated circuits. Today, a chip measuring only a few square millimetres contains kilometres of conductors and billions of electronic switching points. Individual elements of these chips comprise only a few dozen atomic layers (about 10 nm). Here, the connection between technology and quantum mechanics goes in the opposite direction: while with lasers and the like, quantum physical laws protrude into the macro world and thus become technically usable, with computers, technology has developed so far into the quantum world that it would have to fail without knowledge of quantum physical laws.

How the **transistor** got its name: When engineers were looking for a suitable name for their new invention, they considered terms such as semiconductor triode, crystal triode, solid-state triode, iotatron, and numerous other names. In the end, a combination of the words transconductance and resistor won the race.

One example of a quantum phenomenon that is essential knowledge in the fabrication of today's chips is the tunnel effect. Quantum particles can overcome barriers with a certain probability, even if their energy is not sufficient for this according to the laws of classical physics—the particle tunnels through the energy barrier, so to speak. Transferred to our macro world, this would mean that of a thousand footballs kicked against a wall, a handful would reappear on the other side of the wall. Quantum tunnelling is a bizarre property of the microworld that has very real consequences for the construction of computers.

Nuclear fission and the atomic bomb thus only represent the shameful beginning of the translation of knowledge about quantum physical processes into technology. After the abrupt end of the Second World War, a wide variety of applications followed in swift succession. Today, in addition to lasers and computers, the entire field of solid-state physics would be inconceivable without knowledge of quantum effects. Moreover, all the properties of materials such as colour, light transmission, freezing point, magnetism, viscosity, deformability, electrical conductivity, chemical characteristics, and much more, can only be understood using the laws of quantum physics. Again and again, physicists come across surprising macroscopic quantum effects that open the way to exciting technological applications. Graphene, for example, discovered in 2004 by Andre Geim and Konstantin Novoselov, conducts electricity and heat very well and is two hundred times more stable than the strongest steel. This material could make electronics, including computers, orders of magnitude more powerful.

In the Grip of the Military and Industry

The Second World War triggered an enormous mobilisation of scientists and engineers for the military. This was not a temporary effect. Because the sciences had proved essential for modern warfare, this development even intensified after 1945. Support for training scientists and engineers was greatly increased, not least due to military pressure.[3] Between 1945 and 1970, more physicists were trained in the USA, England, and the Soviet Union than during the previous 400 years. In the USA and the USSR in particular, the military became the most important sponsor of research, and the universities lost their autonomy. In the USA, the *Advanced Research Projects Agency* (ARPA, later renamed the *Defense Advanced Research Projects Agency*, DARPA), the branch of the Department of Defense responsible for research funding, pulled all the strings from 1957 onwards. The influence of this lavishly funded organisation extended far beyond the military sphere. In 1969, for example, ARPA developed a computer network called Arpanet, which linked mainframe computers from universities, government agencies, and defence companies across the country. By the mid-1970s, Arpanet included 60 nodes and became the core structure of today's Internet. If we also take into account the historical development of the first computing machines used for nuclear research, it becomes clear that the two core technologies of today's information technology, the Internet and computers, had their beginnings under the auspices of the military.

In addition to the military, industry also began to influence science. For example, the transistor was not developed by an independent university research team, but at Bell Labs, the laboratory of the AT&T telephone company.

The increasing power of the military and industry to determine scientific progress was even a thorn in the side of the outgoing President of the United States, Dwight Eisenhower. In his farewell address to the nation in January 1961, he warned:

> Now this conjunction of an immense military establishment and a large arms industry is new in the American experience. The total influence—economic, political, even spiritual—is felt in every city, every Statehouse, every office of the Federal government. We recognize the imperative need for this development. Yet, we must not fail to comprehend its grave implications. Our toil, resources, and livelihood are all involved. So is the very structure of our society.
>
> In the councils of government, we must guard against the acquisition of unwarranted influence, whether sought or unsought, by the military-industrial complex. The potential for the disastrous rise of misplaced power exists and will persist.

---

[3] R. Abrams, *The US Military and Higher Education: A Brief History*, Annals of the American Academy of Political and Social Science 502(1): 5–28 (1989); D. Kaiser, *History: From blackboards to bombs*, Nature News (28 July 2015); https://www.nature.com/news/history-from-blackboards-to-bombs-1.18056.

But the trend was unstoppable. Science was less and less about a free mind striving for pure knowledge and purposeless cognition. Scientific work and teaching were now evaluated according to whether they served the profit of corporations or state interests—with the military, state, and private industry often closely intertwined. Knowledge became a commodity to which factors such as competition in global rivalry, efficiency, optimisation, innovation, and market orientation apply just as they do to any other product. Terms like "economisation of science", "academic capitalism", "entrepreneurial university", or even "knowledge as a commodity" testify to this. The community of individual scientists and smaller research groups that had a home in universities and private laboratories became Big Science.

The fact is that today, in many fields, research is only possible on an international level and with powerful donors who promise themselves a return on investment. Individual universities could no longer afford large-scale projects such as even more superconducting medical devices or the simulation of our brains by giant computers. The search for third-party funding is now a fundamental part of a university professor's duties.

But what happens when the spirit of research is determined by economic purposes and goals?

- As early as 1935, the Polish bacteriologist and sociologist of science Ludwik Fleck distinguished between two forms of communication in science: on the one hand, within small circles of specialists and, on the other, between scientific specialists and laypersons (i.e., politicians and military personnel, for example). Fleck observed that communication among specialists is more controversial and open-ended than communication which is primarily concerned with producing technologically efficient and commercially attractive applications.
- In 1942, the sociologist Robert K. Merton was alarmed by the willingness of German scientists to place themselves in the service of the Nazi regime. He emphasised that the function of science and its important role in society are endangered when the focus is no longer on gaining knowledge per se, but on the economic benefit of new scientific findings.[4] Today, scientists are faced with the decision of whether to stay at a university, where they will be relatively poorly paid, but relatively free to do research, or to go into industry, where salaries are significantly higher, but the goals are narrowly defined and focused on profitability.

---

[4] R. Merton, *A Note on Science and Democracy*, Journal of Legal and Political Sociology 1: 115 - 126. (1942); R. Merton, *Science and Democratic Social Structure*, Social Theory and Social Structure, pp. 604–15, New York: Free Press edition (original 1949, 1957).

- In 1987, the historian Paul Forman also dealt with this topic.[5] (We encountered his explanation of why quantum physics was born precisely in the post-war Germany of the 1920s in Chap. 8.) He showed that the massive military funding of science after the Second World War not only greatly expanded its scope and importance for society, but also initiated "a qualitative change in its aims and character".
- In this context, it is useful to note that, in a science driven by the prospect of patents or even controlled by politics or the military, the free flow of information between research groups is inhibited. An extreme example from the past shows how self-damaging such censorship can be. In 1937, the German Reich's Minister for Science, Education, and National Education, Bernhard Rust, issued a ban on reading international scientific journals.

The possibly negative influence on scientific methods of the influx of military funds and the shift in focus from fundamental to applied research is still the subject of heated debate today. Most historians of science rely on toned-down versions of Forman's and Merton's theses. Despite the radical changes brought about by military and industrial funding, scientists—at least in the democratic West—still have sufficient autonomy in their quest to understand nature and thus the freedom to pursue the fundamental questions surrounding the functioning of our world. China, however, is taking a completely different path. The country benefits from the freely available results of other countries' basic research and uses them for their translation into technology and thus for a further expansion of its own technological advantage.

History shows that what is actually "purposeless" fundamental science repeatedly gives rise to decisive new technological ideas, often decades later. One example is the mathematical description of the movement of fluids by the Navier–Stokes equations. Since these equations, developed by Claude-Louis Navier (1822) and Sir George Gabriel Stokes (1851), are non-linear and therefore quickly become very complex, they only became important in fluid mechanics, for example, a good 150 years later, when computers and other very specific technological prerequisites were available. A few more decades later, the Navier–Stokes equations found their way into modern

---

[5] P. Forman, *Behind quantum electronics: National security as basis for physical research in the United States, 1940–1960*, Historical Studies in the Physical and Biological Sciences, Vol. 18, Pt. 1, 1987, pp. 149–229 (1987).

weather and climate research.[6] It is quite possible that they will contribute significantly to the survival of humankind.

---

[6] See also: L. Jaeger, *Wege aus der Klimakatastrophe—Wie eine nachhaltige Energie- und Klimapolitik gelingt*, Springer (2021).

12

What is a Human Being?—Our Mind as a Scientifically Ascertainable Entity

The eighty years from 1870 to 1950 were a time of deepest doubts and most ingenious insights. The basic sciences of physics and mathematics emerged strengthened from their greatest crises, while biology went from being a self-sufficient Cinderella to a field that deeply affects the way we live, thanks mainly to the new insights into genetics. But the fantastic novel insights and exciting technologies did not only affect the world we create around us. Knowledge about humans themselves also made enormous progress. Thanks to this, we are closer than ever to answering some of the fundamental questions of philosophy today:

1. Where do we come from?
2. What makes us human?
3. What is consciousness?

Regarding the first question "Where do we come from?", Darwin already had some conclusions to offer. Since the apes living in Africa are most similar to humans, he thought that we must be relatively closely related and have common ancestors there. The heated discussion around this theory created a new field of activity for adventurers in the late nineteenth century: so-called bone hunters searched the world over for the fossil remains of the ancestors of modern humans.

© The Author(s), under exclusive license to Springer Nature
Switzerland AG 2022
L. Jaeger, *The Stumbling Progress of 20th Century Science*,
https://doi.org/10.1007/978-3-031-09618-1_12

- In 1859, almost at the same time as Darwin's first publication of his theory of evolution, Neanderthal man was discovered accidentally near Düsseldorf. More about this relative can be found on the following pages.
- In 1891, the Dutchman Eugène Dubois found a fragment of a skull on the Indonesian island of Java that resembled that of an ape. However, a thigh bone found shortly afterwards definitely pointed to an upright walking creature. Dubois thought he had found a common ancestor of man and ape in the Java Man and presented it to his colleagues with great enthusiasm. But most of them thought the fossil was an ancestor of the apes living today—in their opinion, a creature closely related to man should have had a much larger brain. They were wrong. Today, Java Man is considered to be the first evidence of *Homo erectus* ("the erect man"), a direct ancestor of modern humans, but not of apes.
- In 1925, the physician and anatomist Raymond Dart, who had accepted a teaching position at the University of Johannesburg in South Africa two years earlier, described in the journal Nature the amazingly well-preserved skull of a very young hominid—by then, this was the name given to ancestors whose skeleton suggested an upright gait. He gave the find, which had become famous as the "Taung child", the official name *Australopithecus africanus* ("ape from southern Africa"). He was convinced that this new species was a human ancestor. Again, his colleagues classified the find as a direct ancestor of an ape, because of the small size of the skull. Moreover, it was hard for them to accept that Africa could be the cradle of humankind, even though Darwin had already suggested this. The rejection of Dart's assessment got personal: his former mentor Elliot Smith accused him of having no idea about anatomy, and a second reviewer said Dart's conclusions were ridiculous.
- One of the few scientists who sided with Dart was the Scotsman Robert Broom. The renowned expert in the field of early vertebrates, who like Dart had ended up in South Africa, congratulated him on the important find and decided that he himself would search for early human ancestors. Health problems and professional setbacks meant that he was only able to begin his search in earnest from 1936. By 1948, when Broom was already 81 years old, he and Dart had collected so many fossil specimens that even the sceptics could no longer deny that *Australopithecus* was an ancestor of modern *Homo sapiens*.
- Africa became a hotspot for anthropologists. In 1959, the two British palaeoanthropologists Mary and Louis Leakey found another prehistoric human in the Olduvay Gorge in northern Tanzania, this time with a

distinctive cranial crest. Today, this find is thought to point to a lateral line in the development from *Australopithecus* to *Homo*.

- A year later, one of the Leakeys' sons discovered skull fragments of a more advanced hominid. Tools found near the site indicated that this specimen was already able to use its hands in a very sophisticated way. The researchers named him *Homo habilis* ("skilled craftsman").
- In 1974, anthropologist Donald Johanson found another *Australopithecus* fossil in Ethiopia, which he named Lucy. It was another indication that the members of the *Australopithecus* genus showed very different expressions, from massive-robust to fine-limbed.

The discovery of the Taung baby. Josephine Salmons, one of Raymond Dart's students, discovered a fossilised skull in the house of a family friend in 1923, where it served as a living room decoration. The householder was the director of the Buxton Limeworks quarry, located near the small town of Taung, south-west of Johannesburg. He had taken the fossil home as a curiosity, thinking it was a fossilised monkey skull. Salmons realised that this relic was something special and showed it to Dart, and Dart's research led him to a box of finds collected by a quarry foreman named de Bruyn. This box contained, among other things, the skull of the Taung child, which is one of the most important finds in palaeoanthropology.

These and a large number of other finds produced a picture of bewildering diversity. The idea of a linear evolutionary path that forks into two branches at a certain point, one of which leads via some intermediate stages to modern humans and the other to the apes known today, had to be abandoned. Throughout the entire early history of humankind, several species lived in parallel at the same time, even in directly neighbouring regions. It is only since about 30,000 years ago that *Homo sapiens* has been alone on the globe. The relationships between humans and apes, and especially the precursors of humans, are therefore much more complex than Darwin had imagined.

A Look into the Distant Past

Anthropologists look at and compare their finds very closely to draw conclusions about characteristics such as mode of locomotion, food, and relationships—it is amazing what can be gleaned from a single tooth or a fragment of a finger bone. But dating the fossils is also of the utmost importance in order to trace a family tree. For hundreds of years, researchers were dependent on determining the age of the rock layers in which the fossils were discovered as precisely as possible. This chronological classification is a laborious business, because only when factors such as regional peculiarities of Earth history and climatic conditions are taken into account can a fairly reliable estimate be made. Accompanying fossils or artefacts such as bifaces allow a further narrowing down of the age of the finds. However, the results are always only relative dates, i.e., "older than", "younger than", and "in the same geological period". Absolute dates are only possible to a very limited extent with this method.

From the beginning of the 1950s, researchers had at their disposal a new way of determining age. With radiometric methods, which take into account the natural occurrence of radioactive substances, more precise absolute dates became possible. Here are some examples:

Stratigraphy is the geological discipline that relates the age of rock layers to each other. In the simplest case, the youngest rock layers are near the Earth's surface and the oldest are hidden far below. However, due to geological folding of strata and tectonic influences, the stratigraphic sequences are often more complicated to read. **Bio-stratigraphy** complements the findings: certain fossils only appear after a certain geological age or have become extinct.

- The first step was made by the American Willard Frank Libby, who did research in the field of physical chemistry. With the radiocarbon dating method he developed (^{14}C-method), the age of organic substances can be determined to within a few decades. However, it is not possible to determine the age of finds older than approx. 60,000 years because the ^{14}C concentration reaches the detection limit.
- There are also radioactive isotopes in rocks that decay at a constant rate. For example, ^{235}uranium changes to ^{207}lead with a half-life of 703.8 million years. The higher the proportion of lead in a rock layer compared to the proportion of uranium, the older it is.
- ^{238}Uranium has a half-life of 4.468 billion years and decays to ^{206}lead.

^{14}C method. Due to cosmic radiation, the radioactive carbon isotope ^{14}C, whose atomic nucleus has 8 neutrons instead of 6, is continuously produced in the uppermost layers of the atmosphere. As $^{14}CO_2$, it is absorbed by plants and finally metabolised in all living organisms via the food chain. If the organism dies, it stops taking up ^{14}C. Now the half-life of ^{14}C comes into play. This is 5730 years. Determining the ^{14}C content in the sample thus leads to a very precise age determination.

In theory, Ernest Rutherford had already proposed in 1905 to use the radioactive decay of uranium to determine the age of rock samples. But it was not until the development of the atomic bomb that a sufficient understanding of uranium isotopes and their decay was gained. By 1953, the technique of isotope determination had advanced to the point where the uranium–lead method could be used to determine the age of the Earth as 4.55 billion years ±50 million years.

There are several reasons why, despite this progress, different human phylogenies are still being discussed:

- The time data determined using the various measurement methods available can show large discrepancies, so different research groups evaluate the age of certain finds differently.
- Even though today's mass spectrometers can determine the age of samples weighing only a millionth of a milligram, the calibration curves for isotope decay are still subject to interpretation.
- Fossils with hitherto unknown characteristics are constantly being unearthed during excavations, adding new facets to the history of humankind.

A total of 27 different prehistoric humans have been described so far, as of 2021. It may be that one of the already known species is the last common ancestor of modern humans and apes. Which of them this would be is still unknown. But thanks to another method, scientists are getting closer to unravelling the convoluted paths of human evolution.

The Neanderthal in US

With only 0.5% deviation in the genome, the Neanderthals are our closest relatives. They lived in Europe and Asia between 350,000 and around 30,000 years ago; the oldest findings of *Homo sapiens* unanimously recognised by experts come from north-east Africa and are a good 160,000 years old. The fact that modern humans and Neanderthals populated the Earth at the same time and often in the same places over long periods of time speaks against the long-held assumption that Neanderthals were a direct ancestor of modern humans. So, was Neanderthal man just a side branch in the evolution to *Homo sapiens*, one that could not hold its own in the long run against its cleverer cousin and disappeared forever? Modern DNA analysis proved to be a method that could answer this question.

The founder of palaeogenetics, the Swede Svante Pääbo, was convinced that under favourable circumstances intact DNA could also be obtained from fossils and studied. Contrary to the expectations of experts, he succeeded in 1997 in isolating mitochondrial DNA from a Neanderthal and determining the first sequences. By 2009, 60% of the entire Neanderthal genome was known, and by 2014 it had been completely decoded. In total, he and his team were able to completely determine the genomes of three individuals and publish at least fragments of many other genomes. One of the biggest surprises, however, was their discovery in 2010 that Neanderthal man and *Homo sapiens* produced offspring together! So both the above statements are correct: Neanderthal man is our cousin and at the same time to be found among the ancestors of non-Africans.

Why do people originating from Africa not have Neanderthals as ancestors? Palaeoanthropologists assume that Neanderthals are descended from hominids that made the leap from Africa to Europe and Asia half a million years before *Homo sapiens*. When *Homo sapiens* left Africa for the north, he encountered Neanderthal man, who had long since adapted to living conditions in Europe and Asia. In the genome of modern non-Africans, 1–4% of the genes can be traced back to the Neanderthal, predominantly relating to characteristics of (lighter) skin and (more) hair growth as well as the immune defence system.

DNA analysis of human genomes has allowed a number of other findings:

- The team of Allan Wilson in Berkeley proved as early as 1986, on the basis of mutations in mitochondrial DNA, that *all* women living today have a *single* common ancestor. The "original Eve" lived in Africa 100–200,000 years ago.
- All non-African humans living today are descended from a *Homo sapiens* group of only about 500 individuals who spread out from the area of present-day Ethiopia across the entire globe about 50–60,000 years ago; they were also able to colonise Australia and America via land bridges that still existed at that time.
- The comparison of the human genome, which has been known without gaps since 2003, with the complete version of the chimpanzee genome,

which has been available since 2005, leads to the conclusion that the separation of the two lineages took place around ten million years ago. Both lineages continued to produce reproductive offspring for several million years. The final separation of humans and apes occurred a good seven million years ago.

Neanderthals and Covid-19. Some of the Neanderthal-derived genes in us influence susceptibility to Covid-19 both positively and negatively:
- 30–50% of all non-Africans possess a gene variant on chromosome 12 that is responsible for an improved immune defence against RNA viruses (for example corona and hepatitis C viruses).
- A variation on chromosome 3, on the other hand, increases the probability of contracting life-threatening covid-19 if corona infection occurs.

Today, anthropologists and palaeogeneticists are no longer dependent on being lucky enough to come across bones in which the DNA is still intact due to favourable conservation conditions. DNA can now be extracted from cave sediments, making sites accessible to this method even when their fossils no longer contain sequenceable DNA. Further exciting insights into the human family tree can therefore be expected from these gene analyses.

Knowledge of the human genome goes hand in hand with molecular biological methods that can be used to specifically modify human genetics. This adds a significant sequel to the question "Where do we come from?" For we may now ask "Where are we going?" Indeed, today it is technologically possible to enhance human physical and mental endowments and abilities. Efforts of this kind are not new; dental prostheses and hearing aids have been known since antiquity, and spectacles have assisted human vision since the late Middle Ages. In the twentieth century, technological applications multiplied the options for optimising humans—pacemakers, artificial joints and psychotropic drugs are just a few examples. Direct intervention in our genes opens the door to an even more profound development: the genetic redesign of the human being.

An Organ Steps Out of the Shadows

Now to the second of the three questions posed at the beginning of this chapter: "What makes us human?" A question still fiercely debated among anthropologists today deals with the beginning of human culture. Its essential precondition is language. It is assumed that both the Neanderthals and the second-wave hominids who colonised the rest of the world from Africa were able to communicate through language. The ability to think, which is linked to the ability to speak, probably also proved to be a decisive evolutionary advantage in this conquest.

Thinking and speaking require certain competencies and structures of the brain. But it is precisely this organ that only came into scientific focus very late. The way nerves work also remained unknown for a long time.

- For Aristotle, the brain was a cooling unit for the blood heated by the circulation. He assumed that thinking took place in the heart.
- A few decades after him, the ancient Greek scholars Herophilos and Erasistratos distinguished between nerves that control the body's motor functions and those that guide sensations. The fact that they had no qualms about testing their ideas on slaves, whom they dissected alive, was already met with great astonishment at the time.
- For a long time, dissecting corpses was taboo in Europe for religious reasons. It was not until the Renaissance that scientists began to study anatomy on the human body itself and not just from ancient books. In the sixteenth century, the Flemish physician Andreas Vesalius described the structures of the human brain and corrected numerous errors in ancient teaching.
- Until the nineteenth century, nerves were thought to be hollow like blood vessels and to contain an unknown fluid. The physiologist Emil du Bois-Reymond, mentioned in Chapter five, and also the physicist Hermann von Helmholtz and his assistant Wilhelm Wundt, showed in the 1840s and 1850s that nerves transport, not fluid, but electrical signals between the brain and the body.
- From the second half of the nineteenth century, brain functions were examined experimentally. In 1861, the French physician Paul Broca located the motor speech centre responsible for the grammar of language in the left frontal lobe of the brain. Shortly afterwards, his German colleague Carl Wernicke determined the centre responsible for understanding language in the left temporal lobe. In 1909, the first complete map of the brain was published, linking specific areas with the corresponding mental abilities.
- In 1873, the Italian physician Camillo Golgi accidentally discovered the possibility of staining nerves with silver nitrate salts so that they could be observed in good resolution. Although he was aware of the gaps between the individual nerve cells, he assumed that there was a continuous nerve network whose components could not be isolated from one another.

Throughout his life, he was an opponent of the theory that the network consisted of individual nerve cells.

- The Spanish anatomist Santiago Ramón y Cajal, on the other hand, assumed in his neuron doctrine, which is still recognised today, that the nervous system is a network of specialised cells that are connected to each other as individual building blocks, but not in direct contact. The fact that they nevertheless form a common network is—as was later shown—due to their electrical excitability and conductivity. They selectively transmit electrical impulses across the gaps between them, and process and store them in the network.

- In 1891, the German physician Wilhelm Waldeyer-Hartz, who had already given his name to the chromosome, published the view that the structures visible under the microscope were single, extremely elongated cells. He called them neurons (from the Greek *neúron*: "nerve"). As the smallest units of the nervous system, neurons acquired a similar significance in brain research to atoms in physics and genes in biology.

- Charles Sherrington introduced the term synapse in 1897 for the gap between neurons.

The dispute between **Golgi** and **Cajal** and their respective supporters lasted for decades. Although Cajal was able to convince the world's leading medical researchers of his neuron doctrine at the congress of the German Anatomical Society in Berlin in 1889, Golgi stuck to his opinion that the nervous system was an interconnected network. The fact that they were both awarded the Nobel Prize for Medicine in 1906 probably disturbed both Golgi and Cajal. Cajal wrote about Golgi in his autobiography:

One of the most conceited and self-promoting gifted men I have ever known.

After the basic anatomical relationships in the brain had been clarified, brain research did not pick up speed again until much later. This was largely due to the fact that, thanks to Freud, the psychological unconscious was initially the focus of interest, while later, as will be explained in more detail below, behaviourism investigated the connection between information entering the brain and the resulting behaviour. In 1952, the English physiologists Alan Hodgkin and Andrew Huxley established that the transport of ions, especially sodium, potassium, and calcium ions, is responsible for the transmission of electrical signals in nerve cells. They were awarded the Nobel Prize for Medicine in 1963 for this discovery. But it was not until the 1990s that things really got moving. Neuroscientists call it the decade of the brain. During this period, scientists found out more about this organ and how it works than throughout the whole of history. The period from 2000 to today has also been marked by enormous leaps forward.

A New View of Feeling and Thinking

The way nerves transmit signals is the same throughout the animal kingdom. The structure of brains in different species is also similar. Indeed, the various structures making up the human brain can also be found in all vertebrates, at least in preliminary stages. *Qualitatively*, therefore, there is hardly any difference between human and animal brains; it is the *quantity* of neurons and synapses that makes the human brain so effective and distinguishes it from all other animals.

- Even the 200,000 brain cells of the fruit fly are enough to control flight manoeuvres, find food, and generate offspring.
- Although the brain mass of a dolphin is equal to that of a human, it consists of only 5.8 billion nerve cells, a fraction of the number found in the human brain.
- Chimpanzees have 6.2 billion nerve cells.
- Human brains have around 100 billion nerve cells, each forms up to 200,000 contact points with other nerve cells.

Over time, scientists have succeeded in identifying the structures of the brain and understanding the physiological basis of mental processes better and better. Here is a small selection of the imaging techniques available today, all of which are based on discoveries in subatomic physics:

- Electroencephalography (EEG), which measures voltage fluctuations on the surface of the head, has been available since the early 1930s. Measurement results provide direct information about the electrical activity of different areas of the cerebral cortex; deeper layers are not recorded.
- Magnetic resonance imaging (MRI), developed in the 1970s, can be used to distinguish the finest tissue structures in the brain. The corresponding device generates a strong magnetic field with which the atomic nuclei in the brain align themselves at different rates depending on the type of tissue.
- In positron emission tomography (PET), which has been known since 1975, a weakly radioactive contrast medium is administered to the test

person. When the substance decays, it emits two photons that move in exactly opposite directions. Detectors arranged in a ring around the brain catch these photons and thus detect the exact location of the decay. Since the contrast agent is primarily transported in the body to places with increased metabolism, it can be used to identify the particularly active areas of the brain.

- The fact that oxygenated blood has weak magnetic properties has been known since 1936. Functional magnetic resonance imaging, or (f)MRT for short, developed in the 1980s and 1990s, uses blood flow to depict particularly active areas of the brain in three dimensions and in real time.

With these methods, researchers can watch the brain think. In experiments, the candidates are usually given specific tasks, for example, they may be asked to name colours or shapes. The activity patterns that occur in the brain are measured and the excitation patterns of the nerve cells that occur are brought into a direct relationship with the set task. In this way, the regions in the brain responsible for certain tasks can be localised. In contrast to earlier maps of the brain with clear boundaries, neuroscientists today associate probabilities with the locations of certain functions in the brain. When assigning neural correlates, they divide the brain into tiny spatial units called voxels, the 3D version of two-dimensional pixels. The gradual technical development of imaging methods allows ever more detailed insights into the activity patterns of neurons. However, so far, the resolution is still nowhere near the scale of individual nerve cells.

Thinking About Thinking

The successes in mapping the brain relate almost exclusively to *unconscious* thought processes. In contrast, neuroscience is still in its infancy when it comes to describing the neuronal basis for consciousness, because this level of brain activity, which involves more than the mere reception of sensory stimuli, is apparently not linked to a specific brain structure.

But what is consciousness anyway? We perceive our environment, have thoughts, emotions, and memories—and we *know* that we do all this and can also describe our perceptions. The pioneers of psychology and brain research in the nineteenth century, including Du Bois-Reymond, von Helmholtz, and Wundt, were already interested in consciousness. But with Sigmund Freud, this preoccupation came to a temporary end, because the focus now switched to the unconscious. The question of consciousness was almost banished from the sciences in the first 75 years of the twentieth century. And in the Anglo-Saxon world, which also dominated the global research landscape in psychology and neurology from the 1950s onwards, all hopes were pinned on behaviourism.

It was not until the late 1970s and increasingly since the early 1990s that research turned again to the processes taking place in the brain. Today, however, we are still a long way from being able to describe conscious mental states in detail or even explain them. There is not even a generally accepted definition of consciousness. Philosophers and scientists approach such a definition by distinguishing between different types of consciousness, in contrast to more theological ideas such as the soul:

- **Phenomenal consciousness** refers to our conscious experience of sensations and emotions—for example, pleasure or pain. In a scientific context, such subjective experiences are called *qualia* (singular: quale, from the Latin word *qualis*, meaning "of what kind?"). Put simply, this term describes the idea that perceptions "feel a certain way".
- **Intentional consciousness**, on the other hand, refers to our ability to consciously remember, to play out different scenarios and evaluate them by consciously mentally replaying concrete events, to plan for the future, to have expectations, and much more. We can even think about our thoughts. This awareness enables us to imagine things and to distinguish these subjective ideas from objective facts.
- In addition to phenomenal and intentional consciousness, there is the concept of **self-consciousness**: there is a *self* within us, an *ego* that experiences and thinks. We are aware that it is we ourselves who have phenomenal and intentional consciousness. This ability to think and reflect on ourselves enables us to perceive ourselves as individuals in the world. Human self-consciousness pushed open the door to a cultural evolution that led all human societies to record stories and ask philosophical questions in one form or another. Philosophers also refer to this quality as the human spirit.

Behaviourism treat the brain as a *black box*, assuming that the processes that take place in it cannot be objectively recorded and are therefore irrelevant. Only what is objectively measurable is considered:
- the *input*, i.e., the signals from the outside world that reach the brain,
- the *output*, i.e., the observable behaviour triggered by these signals.

> • In today's science, this reduction of the brain's performance—and thus also of the essence of a human being—to a pure stimulus–response scheme, hardly plays a role any more.

Intentionality refers to goals we pursue with our thinking, but also to the evaluation of actions or perceptions. What we think and what we intend, we can recognise as true or false, good or bad. The concept of intentionality goes back to the German philosopher and psychologist Franz Brentano, who was active in the late nineteenth century.

The basics of sensory experience and thus of phenomenal consciousness, for example fear or pain, are certainly also present in animals. Higher animals obviously also have a mental consciousness. But only humans can detach themselves from themselves mentally and face themselves in thought as in a mirror.

What is known so far about human consciousness? Here are some answers:

- This level of brain activity, which is more than pure thinking, is apparently not linked to a specific brain structure. Not even the assumption that our consciousness is localised in the cerebral cortex can be sustained, for there are also many conscious experiences in which inner, and thus developmentally older, layers of the brain are involved.
- Our consciousness helps us to take in complex, unknown events; for example, we can only grasp a complex sentence or a mathematical problem by consciously reading or listening.
- Consciousness is closely linked to the ability to express thoughts and feelings in language. It is therefore natural to assume that the more highly developed hominids, who already had a language, also had consciousness.

Since conscious mental states represent only a fraction of the processes taking place in the brain, the great challenge is to filter out precisely those neuronal processes that generate conscious states against the ubiquitous background of unconscious information processing. Conversely, the study of people whose brains can only produce limited conscious experience due to injuries and other damage leads to interesting results. It has been shown, for example, that lack of consciousness often results in a massive limitation of phenomenal experiences. In the meantime, such functional limitations of the brain can also be produced on healthy subjects by imposing strong magnetic fields from the outside. This makes it possible to directly investigate the significance of the affected areas for consciousness.

It is still a mystery when, how, and why some neuronal activities lead to conscious experience and others do not. However, there have been some first partial successes. One example concerns the phenomenon of *binocular rivalry*. It occurs when the two eyes of a subject are presented with two different images that cannot be integrated into a unified image, for example, a horizontal red bar is presented to one eye and a vertical green bar to the other. It can be shown that both pieces of information reach the brain via the neurons. Nevertheless, the person consciously perceives only a red horizontal bar *or* a green vertical bar, never both together. Moreover, the person cannot control the changes in the perception of the red and green bar voluntarily. In this context, groups of neurons can be identified that are active exactly when the spontaneous change takes place. It may be that these groups determine which image reaches consciousness.

At the Limits of the Explainable

Many philosophers believe that it is impossible in principle to grasp the essence of consciousness scientifically. A number of arguments support this view. Here is a selection:

- The **subjectivity of consciousness** provides the essential justification for this scepticism. Natural science deals with phenomena that occur independently of the observer. It is precisely this objectivity that is a precondition for scientific work. But consciousness takes place in subjective worlds. It is impossible to grasp it with the methods of science, which invokes objective regularities.

- The so-called **qualia problem** is that there is no recognisable connection between measurable neuronal states and the subjective sensations of experience—the qualia. On a purely material level, it can be explained without any gaps how stimuli are conducted to the brain and processed there, and also how they cause a physical reaction. But conscious, subjective experience cannot be explained from these neuronal states.[1]

---

[1] P. Bieri, *What makes consciousness an enigma?*, in: W. Singer (ed.); *Gehirn und Bewusstsein*, Heidelberg, Spektrum. (1994) pp.172–180; *Consciousness. Contributions from Contemporary Philosophy*, Thomas Metzinger (1996) pp. 61–77, and Spektrum der Wissenschaft 10/1992, pp. 48–56.

- The **intentionality problem** goes beyond the qualia problem and refers to the consciousness associated with an intention: thoughts can be true or false, sensations can have a concrete cause or not. Nerve cells and the neuronal states they trigger, on the other hand, know neither truth criteria nor conscious goals; their activity follows only the known laws of nature: either they are stimulated or not, but whether they are stimulated "for a good reason" or not is not a valid category for them.

> The **subjectivity of consciousness** also includes the fact that it is not possible for us to fully and objectively comprehend the consciousness of another person. Nor can we imagine the subjective world of perception of an animal. The American philosopher Thomas Nagel made a particularly impressive point of this in his 1974 article *What is it like to be a bat?*[1]: we definitely cannot imagine, let alone comprehend, the feelings of a bat.

The qualia and intentionality problems are similar to the mind–body problem formulated in the seventeenth century by the French philosopher René Descartes: How are mind and soul connected to the material body? Descartes introduced the view that simple brain functions such as muscle control were to be separated from higher mental tasks such as thinking and self-awareness. He saw the latter not as activities of the brain but as expressions of the immaterial soul. Today, more and more mental functions can be traced back to concrete neuronal states, even those which for Descartes clearly belonged to the competence of the soul. For example, he had classified memory and recall as immaterial. For these abilities, however, neuroscience has discovered convincing correlations in recent years that point to purely material processes. The existence of nerve groups that become active during the conscious process of binocular rivalry also speaks against Descartes. But Descartes' separation into a material and a "mental" part of brain performance has not yet been completely refuted: the question of whether *all* consciousness will ultimately be explainable on a purely material level is still open.

Most scientists today assume that everything in nature, including our *consciousness*, can be explained completely and in all its facets by physical laws, without having to take spiritual factors into account. This idea is not new; it already appeared in antiquity, for example in Democritus and Epicurus. There are reasons for this view held by brain researchers:

---

[1] Th. Nagel, *What Is It Like to Be a Bat?* The Philosophical Review 83 (4): 435–450 (1974).

- They refute the sceptics' main argument—the impossibility of describing subjective processes objectively—with the objection that the processes in the quantum world are also observer-dependent and thus not objective. Nevertheless, science has successfully explored the subatomic world, uncovered correlations and translated the findings into ubiquitous technologies.
- In the history of natural science, questions have often been considered inaccessible to science, but later proved to be precisely answerable by science. Examples include the history of human origins, the atomic world, and the structure and development of the universe.

Another reason for this optimism lies in the assumption that consciousness is a phenomenon of emergence: the material, energetic, and functional processes in the brain are sufficiently complex for consciousness to emerge from them as an autonomous behaviour. Two examples of emergence from other fields of science:

- Individual atoms have neither pressure nor temperature, nor are they in any particular state of aggregation. Only when many atoms come together do these properties arise.
- In chaos research, the term emergence refers to the self-organised emergence of ordered structures out of disorder.

Since cases of emergence have also been shown to be explainable and calculable in other areas of science, neuroscientists hope that consciousness can also be explained and understood in principle, even if the processes underlying human consciousness are probably far more complex than the examples of emergence discovered so far.

Emergence is the experience that the whole can be more than the sum of its parts. In this case, consciousness is more than the mere interaction of neurons. This phenomenon goes beyond the explanatory patterns of the *classical* natural sciences, in which the whole can always be broken down analytically into individual phenomena, and individual phenomena can be put together in a synthesis to form a whole. Modern physics, chemistry, and biology involve many emergent phenomena.

Emergence also provides an explanation for the fact that humans are probably the only living beings to have developed consciousness. The human brain is probably the only one in the animal kingdom that reaches the level of complexity necessary for the emergence of emergent consciousness, with its billion nerve cell interconnections.

For many philosophers, on the other hand, the search for an explanation for consciousness will forever remain an impossible scientific question to answer. The majority of them are of the opinion that the objectivity of a possible explanation and the subjectivity of the object of investigation (as well as the executive organ—consciousness) can never be brought together. For them, on closer inspection, it is not at all clear where exactly the riddle of consciousness lies, and thus it is not clear how convincing solutions to the riddle could be discerned. The Swiss philosopher and writer Peter Bieri puts it this way:

> Something applies to the riddle of consciousness that does not apply to other riddles: we have no idea what would count as a solution, as understanding.

Step by Step

The arguments of the sceptics cannot simply be pushed aside. Moreover, the examples listed above of areas of science that were once considered "impossible" but later produced significant findings have not been fully clarified.

- The human family tree still holds many mysteries.
- A very complex quantum theory has been worked out for the atomic and subatomic world, which is not yet complete.
- There are still many unanswered questions about the history of the universe.

But apart from objectivity, this is also part of the "genetics" of the sciences: the tireless effort to track down the riddles of life and the world, even if the obstacles seem almost insurmountable. Neuroscientists are giving special priority to the search for the neuronal correlates of conscious perception, i.e., phenomenal consciousness, and thus to solving the qualia problem. They have already compiled many findings that explain the purely functional phenomena of seeing, hearing, smelling, tasting, and touching neurologically. The research effort is also on the trail of the functioning of other cognitive phenomena such as learning, memory, and problem solving. There is a direct, measurable connection between all these brain performances and locally bounded activities of neurons.

But successes like these only touch the periphery of the qualia problem, the real questions remain unanswered.

- How can the connection between subjective and objective states in consciousness be explained? The Australian philosopher David Chalmers calls this question the "hard problem of consciousness".
- How is consciousness and subjective experience possible in an objective, physical universe? Or put another way, how can it be that we subjectively experience objective sensory perceptions—the nerve cell is stimulated or not—as a conscious ego? (Du Bois-Raymond had already asked exactly this question in 1872 in the sense of Ignorabimus as the fifth of seven world puzzles, as we saw in Chap. 5).
- Why don't humans simply perceive stimuli without consciousness, just like the insects?

Not only does the qualia problem elude our understanding, but there are even fewer answers to the intentionality problem and in particular the question of the origin of self-consciousness.

- How is it that we perceive ourselves as a conscious, subjectively cognizing "I" that is at the centre of an objective universe? Who or what is this entity called "I" that has subjective experiences and intentions?
- What causes us to recognise our experience as a unity among all the many sensory impressions, i.e., that there is a coherence of consciousness? On closer examination, a person's knowledge that he or she is always the same individual is one of the brain's most amazing achievements. How does it realise that the myriad data reaching it from the sense organs concerns "itself"? Even the fact that the individual neurons of the nervous system are replaced by new cells in rotation does not change the fact that the brain perceives itself as a coherent entity.

It should surprise no one that for the German–American neuroscientist Christof Koch, who developed the theory of neuronal correlates together with Francis Crick, human consciousness is one of the most puzzling characteristics of the entire universe.

The Self-Model in the World Model

As in a complicated dance, philosophy and neuroscience simultaneously contradict and complement each other and drive each other to peak performance. One of the controversially discussed topics is the relationship between the inner and outer world in perception. Immanuel Kant's philosophy already contained at its core the statement that we cannot experience an external world that is independent of us. Many of today's philosophers go one step further. They assume that the perceived world is a sophisticated illusion that is constantly recreated by our brain, selectively mapping and processing information.[3] We *believe* that the world is as our brain represents it, but in fact all our perceptions take place *within ourselves*, interacting with what is outside of us. By providing us with a projection of the outside world, our brain helps us to process information as efficiently as possible, to make appropriate predictions and to interact socially—these are precisely the evolutionarily evolved abilities that have made humans the most successful species of all time.

This philosophical theory can be investigated experimentally by connecting the conscious self-model in our brain via brain–computer interfaces with external systems, for example robots or virtual bodies (avatars). Exciting new insights are already emerging here.[4]

> Philosophers talk about our brain providing us with a **model of the world.** In the real world, there are no colours, sounds, etc. Our ego occurs as part of this world model, called the **self-model**.

The currently prevailing view of most neuroscientists can be summarised as follows:

- Our consciousness is a mental world in our brain within which we think, perceive, feel, and plan with a sense of self.
- Conscious experience is a representation in our brain that creates a model of reality for us. This biological data format presents information about the world in a particular way that focuses our attention, creates an internal representation of the world, and ultimately enables us to simulate different temporal scenarios.

---

[3] Particularly beautifully presented in the book by T. Metzinger, *The Ego Tunnel—The Science of the Mind and the Myth of the Self,* New York: Basic Books (2010).

[4] For more details see: https://larsjaeger.ch/virtual-reality-and-consciousness-technologies-the-new-pos sibilities-of-brain-computer-interfaces/.

- The consciousness of ourselves, our ego, first manifests itself in subjective perceptions and feelings.
- At the same time, it is an experience-constituting component of our world—without ego, no experience. That is why it is so difficult to grasp our consciousness objectively, that is, from the outside.
- The debate about whether quantum behaviour play a role in consciousness is still ongoing.

Can man's subjective consciousness be explained in an objective universe? Or does it represent a final mystery, an eternal blind spot on the map of the scientific world view? Without an answer to this question, the central question of all philosophies also remains open:

What is man?

Epilogue: The Fifth Virtue of Science

When, from the seventeenth century onwards, modern science, as the initial spark of the Enlightenment, began to replace religious explanations of the world, this turned all previously valid patterns of thought on their head. The prerequisites for this triumphant advance of the natural sciences were four virtues[1]:

- **Turning away from dogmas** requires an uncompromisingly open attitude towards one's own knowledge. In the beginning, it was a matter of breaking away from many ancient ideas and religious patterns of thought. Later, it became important to rethink established views in science without reservation, when new facts demanded it.
- **Confidence in one's own observations** enabled release from the power of dogma. Today, the individual curiosity of scientists still leads to their own observations, which either confirm or question the generally accepted truths.
- **Trust in incorruptible mathematics** accompanies scientists in their search for knowledge. It is the incorruptible language of the laws of nature and serves as the basis of our understanding.
- **The application of knowledge for the benefit of humanity**. Today's technological living conditions in developed countries are almost paradisiacal compared to those that prevailed just a few generations ago.

---

[1] See L. Jaeger, (only in German) *Sternstunden der Wissenschaften—Eine Erfolgsgeschichte des Denkens*, Südverlag (2020).

These four virtues provide the basis and condition for scientific thought and action. The first three are firmly anchored in the sciences, while there is undoubtedly still a need to catch up with the fourth virtue.

Even to this day, the Church has not recovered from the turning point of the Enlightenment and the subsequent decline of its interpretive sovereignty. However, the fact that modern science was also led to the edge of the abyss by a deep crisis, but retook control within a few decades, is hardly known. Around 1900, scientists had reached their limits, because at that time, in order to progress further, they had to leave the realm of the immediately visible. Within a very short period of time, many of the certainties they had hitherto believed to be unquestionable collapsed. Only with the help of a fifth virtue were they able to enter completely new territory and thus overcome this crisis: **intuitive genius**.

This brought the necessary creativity into play without betraying sober rationality.

The history of science over the past 150 years shows that, just as a great leap in knowledge is about to be taken, there are almost always intuitive, sometimes even irrational ideas put forward by individual scientists. It can thus be understood as a dialectical process between occasional outbursts of genius and constant, soberly rational diligence in thinking and observing. In this way, intuitive genius became a particularly strong driver for the decisive breakthroughs of the twentieth century.

In science, such ingenious, overflowing intuition is controlled by methodological doubt and the arbitration of empirical experience. Theories born of intuition that cannot ultimately be measured against clear and irrefutable facts soon disappear from the scene. *Ingenious intuition* in science also includes the objective quality of thoughts. Intuition, on the other hand, which is experiencing a new heyday in today's society, is of a completely different kind: as a *non-genius intuition,* it appeals solely to a gut feeling. Its actors do not want to know *how the world is,* but they see the world as *they want to imagine it.* They do not hesitate to resort to alternative facts to underpin their sometimes intellectually limited, often straightforwardly stupid thoughts.

The path of science from dogma and superstition to rational thinking and empirical research has been and still is long and arduous. All the more valuable is the realisation that the search for knowledge never ends. This was a (new) scientific characteristic recognised by the philosopher Karl Popper: Popper argued in the 1920s that logically no scientific theory can be confirmed once and for all from an experimental test, whereas a single experimental counterexample can logically disprove it. Popper's presentation of the logical asymmetry between verifiability and falsifiability is at the heart of his theory of science, which is still known today. De facto, according to him, a theory can only be considered scientific if it is (potentially) falsifiable. This is also a revolutionary move within the sciences themselves, since until the late nineteenth century the opinion of physicists had been that theories, once confirmed and consistent, were valid forever.

Many scientific laypeople, on the other hand, wish for immutable truths. At a time when populists abuse this longing for their own purposes, when an irrational criticism of science is on its way to becoming socially acceptable, and when doubts about it are being presented ever more aggressively, we must consistently ensure that the voice of rationality remains clearly heard.

References

Abrams, R. (1989). The US military and higher education: A brief history. *Annals of the American Academy of Political and Social Science, 502*(1).

Bateson, W. (1902). *Mendel's principles of heredity*. Cambridge University Press.

Bayes, T. (1908). *Versuch zur Lösung eines Problems der Wahrscheinlichkeitsrechnung*. Verlag von Wilhelm Engelmann.

Bell, J. (1987). *Speakable and unspeakable in quantum mechanics*. Cambridge University Press.

Blair, C. (25 February, 1957). *Passing of a great mind, Life*, pp. 89–104; also at https://books.google.ch/books?id=rEEEAAAAMBAJ&pg=PA89&redir_esc=y#v=onepage&q&f=false

Bohr, N. (1958). *Atomic physics and human knowledge*. John Wiley and Sons, New York.

Boltzmann, L. (1908). *Lecture on Maxwell's theory of electricity and light*. Barth.

Borel, E. (1898). *Leçons sur la théorie des fonctions*. Gauthier Villars.

Born, M. (1957). *Physik im Wandel meiner Zeit*. Vieweg.

Butterfield, H. (1997, first edition 1957). *The origin of modern science*. The Free Press, Revised edition.

Caves, C., Fuchs, C., & Schack, R. (2002). Quantum probabilities as Bayesian probabilities. *Physical Review A, 65*, 022305.

Darwin, C. (1859). *The origin of species*. John Murray.

Darwin, C. (1871). *The descent of man and sexual selection*. John Murray.

Darwin, C. (1993). *The autobiography of Charles Darwin: 1809–1882*. WW Norton & Co; Revised edition.

de Vries, H. (1889). *Intracellular pangenesis*. G. Fisher.

de Vries, H. (1901). *Die Mutationstheorie*. Veit & comp.

© The Editor(s) (if applicable) and The Author(s), under exclusive
license to Springer Nature Switzerland AG 2022
L. Jaeger, *The Stumbling Progress of 20th Century Science*,
https://doi.org/10.1007/978-3-031-09618-1

Dirac, P. (1927). The quantum theory of the emission and absorption of radiation. *Proceedings of the Royal Society of London A, 114*, 243–265.

Dirac, P. (1 February, 1928). The quantum theory of the electron. *Proceedings of the Royal Society, London.*

Dyson, G. (2012). *Turing's cathedral: The origins of the digital universe.* Vintage.

Eckart, W. (2005). *History of medicine.* Springer.

Einstein, A. (2007). In A. Calaprice (Ed.), *The ultimate quotable Einstein.* Princeton University Press.

Einstein, A., Podolsky, B., & Rosen, N. (15 May, 1935). Can quantum-mechanical description of physical reality be considered complete? *Physical Review, 47*, 777.

Feynman, R. (1961). *The Feynman lectures on physics,* vol. I, Reading, Mass, Addison-Wesley Pub. Co.

Feynman, R. (1985). *Surely you're joking, Mr. Feynman!—Adventures of a curious physicist.* Piper Verlag (originally published by W. W. Norton Verlag, New York). *Surely you're joking, Mr. Feynman! adventures of a curious character*

Feynman, R. (1990). *QED.* Princeton University Press, Princeton.

Feynman, R. (2005). *The pleasure of finding things out.* Basic Books.

Foreman, P. (1971). Weimar culture, causality, and quantum theory, 1918–1927: Adaptation by German physicists and mathematicians to a hostile intellectual environment. *Historical Studies in the Physical Sciences, 3.*

Foreman, P. (1987). Behind quantum electronics: National security as basis for physical research in the United States, 1940–1960. *Historical Studies in the Physical and Biological Sciences, 18*(1), 149–229.

Freud, S. (1899). *The interpretation of dreams.* Franz Deuticke.

Friebe, C., et al. (2015). *Philosophy of quantum physics.* Springer.

Good, I. J. (Ed.). (1962). *The scientist speculates.* Heinemann & Basic Books.

Haber, L. F. (1986). *The poisonous cloud. Chemical warfare in the first world war.* Oxford.

Heisenberg, W. (1934). *Wandlungen der Grundlagen der exakten Naturwissenschaft in jüngster Zeit.* Lecture to the Society of German Natural Scientists and Physicians, Hanover, 17 September 1934, Angewandte Chemie 47.

Heisenberg, W. (1971). *Physics and beyond: Encounters and conversations.* Harper & Row.

Heisenberg, W. (1974). *Across the frontiers.* Harper & Row.

Hermann, G. (1935). Die naturphilosophischen Grundlagen der Quantenmechanik, §7. *Abhandlungen der Fries'schen Schule (ASFNF), 6*(2)

Hilbert, D. (1930a). Naturerkennen und Logik, Naturwissenschaften. Published in: *Gesammelte Abhandlungen* (vol. 3, p. 378). Available at (in German): https://www.rschr.de/Htm/David_Hilbert_Naturerkennen_und_Logik.htm

Hilbert, D. (1930b). Naturerkennen und Logik. Published in: *Gesammelte Abhandlungen* (vol. 3, p. 378). Available at (in German): https://www.rschr.de/Htm/David_Hilbert_Naturerkennen_und_Logik.htm

Hoyer, U. (1984). Ludwig Boltzmann und das Grundlagenproblem der Quantentheorie. *Zeitschrift Für Allgemeine Wissenschaftstheorie, 15*(2), 201–210.

Huxley, T. (1863). Ueber die Beziehungen des Menschen zu den nächstniederen Thiere. In *Zeugnisse für die Stellung des Menschen in der Natur*. Three treatises, transl. by J. Victor Carus, Vieweg, Braunschweig.

Jaeger, L. (2015). *Die naturwissenschaften—ein biographie*. Springer.

Jaeger, L. (2016). *Wissenschaft und Spiritualität—Universum, Leben, Geist, Zwei Wege zu den großen Geheimnissen*. Springer.

Jaeger, L. (2017). *Supermacht Wissenschaft—Unsere Zukunft zwischen Himmel und Hölle*. Gütersloher Verlagshaus.

Jaeger, L. (2018). *The second quantum revolution*. Springer.

Jaeger, L. (2019). *Mehr Zukunft wagen—Wie wir alle vom Fortschritt profitieren*. Gütersloher Verlagshaus.

Jaeger, L. (2020a). An old promise of physics—Are we moving closer towards controlled nuclear fusion? *International Journal for Nuclear Power, 65*(11/12).

Jaeger, L. (2020b). *Sternstunden der Wissenschaften—Eine Erfolgsgeschichte des Denkens*. Südverlag.

Jaeger, L. (2021). *Ways out of the climate catastrophe. Ingredients for a sustainable energy and climate policy*. Springer.

Jammer, M. (1995). *Einstein und die Religion*. Universitätsverlag Konstanz.

Kaiser, D. (28 July, 2015). *History: From blackboards to bombs*. Nature News. https://www.nature.com/news/history-from-blackboards-to-bombs-1.18056

Keynes, J. M. (1963). Economic possibilities for our grandchildren, (1930), In *Essays in persuasion*. W.W. Norton & Co.

Kunmar, M. (2008). *Quanten—Einstein*. Berlin Verlag.

Laplace, P. -S. (1814). *Philosophischer Versuch über die Wahrscheinlichkeiten*. Verlag Harri Deutsch (2003) (Current edition).

Laughlin, R. (2007). *A different universe: Reinventing physics from the bottom down*. Basic Books.

Levene, P., & London, E. S. (1929). The structure of thymonucleic acid. *Journal of Biological Chemistry, 83*

Milburn, G. J. (1997). *Schrodinger's machines: The quantum technology reshaping everyday life*. W. H. Freeman.

Morgan, T. (1915). *The mechanism of Mendelian heredity*. Henry Holt and Company.

Morgan, T. (September 1917). The theory of the gene. *The American Naturalist, 51*(609), 513–544 (32 pages).

Mermin, N. D. (April 1989). What's wrong with this pillow? *Physics Today*

Merton, R. (1942). A note on science and democracy. *Journal of Legal and Political Sociology, 1*, 115–126.

Merton, R. (Original 1949, 1957). Science and democratic social structure, In *Social theory and social structure* (pp. 604–615). Free Press.

Metzinger, T. (2010). *The ego tunnel—the science of the mind and the myth of the self*. Basic Books.

Nagel, T. (1974). What is it like to be a bat? *The Philosophical Review, 83*(4), 435–450.

Petković, M. (2009). *Famous puzzles of great mathematicians*. American Mathematical Society. Also at https://archive.org/details/famouspuzzlesgre00mpet/page/n175/mode/2up

Schmaltz, F. (2005). *Kampfstoff-Forschung im Nationalsozialismus*. Wallstein.

Schrödinger, E. (1935). Discussion of probability relations between separate systems. *Proceedings of the Cambridge Physical Society, 31,* 55.

Schrödinger, E. (1944). *What is life?* Cambridge University Press.

Szöllösi-Janze, M. (1998). *Fritz Haber 1868–1934*. Beck.

Turing, A. (1937). On Computable Numbers, with an Application to the Decision Problem. *Proceedings of the London Mathematical Society, London*

Von Neumann, J. (1932). *The mathematical foundations of quantum mechanics*. Springer.

Von Neumann, J. (1951). The general and logical theory of automata. In L. A. Jeffress (Ed.), *Cerebral mechanisms in behaviour; the Hixon Symposium* (pp. 1–41). Wiley.

von Weizsäcker, C. F. (1981). *A look at plato—Theory of ideas, logic and physics*. Reclam.

von Weizsäcker, C. F. (2004). *Pioneer of physics, philosophy, religion, politics and peace research* (SpringerBriefs on Pioneers in Science and Practice). Springer.

Watson, J., & Crick, F. (25 April, 1953). Molecular structure of nucleic acids: A structure for deoxyribose nucleic acid. *Nature, 171,* 737–738.

Watson, J. (1968). *The double helix: A personal account of the discovery of the structure of DNA*. Touchstone (does not publish any longer).

Weyl, H. (1968). In K. Chandrasekharan (Ed.), *Gesammelte Abhandlungen* (vol. IV, pp. 651–654). Springer-Verlag. Also available at (in German only): https://www.spektrum.de/wissen/hermann-weyl-1885-1955-mathematischer-universalgelehrter/1371662

Wigner, E. (1983). Remarks on the mind–body question. In J. Mehra (Ed.), *Philosophical reflections and syntheses* (pp. 247–260), Springer.

Index

Printed in the United States
by Baker & Taylor Publisher Services